AF696476

Traumatic Injury Research at NIOSH

Reviews of Research Programs of the National Institute for Occupational Safety and Health

Committee to Review the NIOSH Traumatic Injury Research Program

Board on Population Health and Public Health Practice

INSTITUTE OF MEDICINE *AND*
NATIONAL RESEARCH COUNCIL
OF THE NATIONAL ACADEMIES

THE NATIONAL ACADEMIES PRESS
Washington, D.C.
www.nap.edu

THE NATIONAL ACADEMIES PRESS • 500 Fifth Street, N.W. • Washington, DC 20001

NOTICE: The project that is the subject of this report was approved by the Governing Board of the National Research Council, whose members are drawn from the councils of the National Academy of Sciences, the National Academy of Engineering, and the Institute of Medicine. The members of the committee responsible for the report were chosen for their special competences and with regard for appropriate balance.

This study was requested by the National Institute for Occupational Safety and Health of the Centers for Disease Control and Prevention and supported by Award No. 211-2006-19152, T.O. #1, between the National Academy of Sciences and the Centers for Disease Control and Prevention. Any opinions, findings, conclusions, or recommendations expressed in this publication are those of the author(s) and do not necessarily reflect the view of the organizations or agencies that provided support for this project.

International Standard Book Number-13: 978-0-309-12507-9
International Standard Book Number-10: 0-309-12507-3

Additional copies of this report are available from the National Academies Press, 500 Fifth Street, N.W., Lockbox 285, Washington, DC 20055; (800) 624-6242 or (202) 334-3313 (in the Washington metropolitan area); Internet, http://www.nap.edu.

For more information about the Institute of Medicine, visit the IOM home page at: **www.iom.edu.**

Copyright 2009 by the National Academy of Sciences. All rights reserved.

Printed in the United States of America

Cover credit: Photos reprinted with permission from Abe Martinez and Don Pollard, 2008.

Suggested Citation: IOM (Institute of Medicine) and National Research Council. 2009. *Traumatic Injury Research at NIOSH.* Committee to Review the NIOSH Traumatic Injury Research Program. Rpt. No. 6, Reviews of Research Programs of the National Institute for Occupational Safety and Health. Washington, DC: The National Academies Press.

THE NATIONAL ACADEMIES

Advisers to the Nation on Science, Engineering, and Medicine

The **National Academy of Sciences** is a private, nonprofit, self-perpetuating society of distinguished scholars engaged in scientific and engineering research, dedicated to the furtherance of science and technology and to their use for the general welfare. Upon the authority of the charter granted to it by the Congress in 1863, the Academy has a mandate that requires it to advise the federal government on scientific and technical matters. Dr. Ralph J. Cicerone is president of the National Academy of Sciences.

The **National Academy of Engineering** was established in 1964, under the charter of the National Academy of Sciences, as a parallel organization of outstanding engineers. It is autonomous in its administration and in the selection of its members, sharing with the National Academy of Sciences the responsibility for advising the federal government. The National Academy of Engineering also sponsors engineering programs aimed at meeting national needs, encourages education and research, and recognizes the superior achievements of engineers. Dr. Charles M. Vest is president of the National Academy of Engineering.

The **Institute of Medicine** was established in 1970 by the National Academy of Sciences to secure the services of eminent members of appropriate professions in the examination of policy matters pertaining to the health of the public. The Institute acts under the responsibility given to the National Academy of Sciences by its congressional charter to be an adviser to the federal government and, upon its own initiative, to identify issues of medical care, research, and education. Dr. Harvey V. Fineberg is president of the Institute of Medicine.

The **National Research Council** was organized by the National Academy of Sciences in 1916 to associate the broad community of science and technology with the Academy's purposes of furthering knowledge and advising the federal government. Functioning in accordance with general policies determined by the Academy, the Council has become the principal operating agency of both the National Academy of Sciences and the National Academy of Engineering in providing services to the government, the public, and the scientific and engineering communities. The Council is administered jointly by both Academies and the Institute of Medicine. Dr. Ralph J. Cicerone and Dr. Charles M. Vest are chair and vice chair, respectively, of the National Research Council.

www.national-academies.org

COMMITTEE TO REVIEW THE NIOSH TRAUMATIC INJURY RESEARCH PROGRAM

BRIAN L. STROM (*Chair*), Professor and Chair, Center for Clinical Epidemiology and Biostatistics, University of Pennsylvania School of Medicine, Philadelphia, PA

ROBIN BAKER, Director, Labor Occupational Health Program, School of Public Health, University of California at Berkeley

LESLIE I. BODEN, Associate Chair and Professor, Department of Environmental Health, Boston University School of Medicine, Boston, MA

BARRY BOZEMAN, Ander Crenshaw Professor of Public Policy and Regents' Professor of Public Policy, State Data and Research Center, Department of Public Administration and Policy, University of Georgia School of Public Health and International Affairs, Athens

STEPHEN W. HARGARTEN, Professor and Chair, Department of Emergency Medicine, Director, Injury Research Center, Wisconsin Injury Research Center, Medical College of Wisconsin, Milwaukee

BRIAN M. KLEINER, Professor, Grado Department of Industrial and Systems Engineering, Virginia Polytechnic Institute and State University, Blacksburg

TOM B. LEAMON, Director Emeritus, Liberty Mutual Research Center for Safety and Health, Vice President, Liberty Mutual Insurance Company, Hopkinton, MA

JAMES M. MELIUS, Director, New York State Laborers' Health and Safety Fund, Safety Trust Fund, Albany

MARK S. REDFERN, Vice Chair and Undergraduate Program Professor, Departments of Bioengineering, Otolaryngology, and Industrial Engineering, University of Pittsburgh School of Engineering, Pittsburgh, PA

GORDON R. REEVE, Corporate Epidemiologist (retired), Occupational Health and Safety, Ford Motor Company, Canton, MI (resigned from the committee May 4, 2007)

JOSEPH J. SCHWERHA, Professor and Director, Environmental Medicine Residency, Occupational and Disaster Preparedness Programs, Graduate School of Public Health, University of Pittsburgh School of Public Health, Pittsburgh, PA

Framework Committee Liaison

LETITIA K. DAVIS, Director, Occupational Health Surveillance Program, Massachusetts Department of Public Health, Boston

Project Staff

KATHLEEN STRATTON, Study Director
MORGAN A. FORD, Program Officer
KRISTINA VAN DOREN-SHULKIN, Senior Program Assistant

Independent Report Reviewers

This report has been reviewed in draft form by individuals chosen for their diverse perspectives and technical expertise, in accordance with procedures approved by the National Research Council's Report Review Committee. The purpose of this independent review is to provide candid and critical comments that will assist the institution in making its published report as sound as possible and to ensure that the report meets institutional standards for objectivity, evidence, and responsiveness to the study charge. The review comments and draft manuscript remain confidential to protect the integrity of the deliberative process. We wish to thank the following individuals for their review of this report:

John C. Bailar III, Department of Health Studies, University of Chicago
Marion Gillen, Center for Occupational and Environmental Health, University of California at Berkeley
Fernando A. Guerra, San Antonio Metropolitan Health District
Guohua Li, Department of Anesthesiology, College of Physicians and Surgeons, Columbia University
Barbara Materna, Occupational Health Branch, California Department of Health Services
Franklin E. Mirer, School of Health Sciences, Hunter College of the City University of New York
Corinne Peek-Asa, Injury Prevention Research Center, University of Iowa
Peter Rosen, Harvard University Medical School

Gordon S. Smith, Department of Epidemiology and Preventive Medicine, University of Maryland
Lorann Stallones, Colorado Injury Control Research Center, Department of Psychology, Colorado State University

Although the reviewers listed above have provided many constructive comments and suggestions, they were not asked to endorse the conclusions or recommendations nor did they see the final draft of the report before its release. The review of this report was overseen by **Enriqueta C. Bond**, Burroughs Wellcome Fund, and **Michael I. Posner**, University of Oregon. Appointed by the National Research Council and Institute of Medicine, they were responsible for making certain that an independent examination of this report was carried out in accordance with institutional procedures and that all review comments were carefully considered. Responsibility for the final content of this report rests entirely with the authoring committee and the institution.

Contents

Tables, Figures, and Boxes

TABLES

FIGURES

BOXES

Abbreviations and Acronyms

AFS	Alaska Field Station
AKDOL	Alaska Department of Labor
ANSI	American National Standards Institute
AOISS	Alaska Occupational Injury Surveillance System
ASABE	American Society of Agricultural and Biological Engineers
ATR	Alaska Trauma Registry
AutoROPS	auto-deploying rollover protective structure
BLS	U.S. Bureau of Labor Statistics
CAIS	Childhood Agricultural Injury Survey
CARE	Children's Act for Responsible Employment
CDC	Centers for Disease Control and Prevention
CFIT	controlled flight into terrain
CFIVSA	Commercial Fishing Industry Vessel Safety Act
CFOI	Census of Fatal Occupational Injuries
CFR	Code of Federal Regulations
CROPS	cost-effective rollover protective structure
DOL	U.S. Department of Labor
DSR	Division of Safety Research (NIOSH)

EMS	emergency medical services
ERC	Education and Research Center
ESA	Employment Standards Administration
FAA	Federal Aviation Administration
FACE	Fatality Assessment and Control Evaluation (Program)
FFFIPP	Fire Fighter Fatality Investigation and Prevention Program
FISH	Fishing Industry Safety and Health
FLSA	Fair Labor Standards Act
FTE	full-time equivalent
FY	fiscal year
HEROES	Homeland Emergency Response Operational Equipment Systems
HHS	U.S. Department of Health and Human Services
HO	Hazardous Occupations Order
IAFF	International Association of Fire Fighters
IAWG	Interagency Working Group
IFISH	International Fishing Safety and Health
IOM	Institute of Medicine
ITCP	internal traffic control plan
JAMA	*Journal of the American Medical Association*
M-CAIS	Minority Farm Operator Childhood Agricultural Injury Survey
NASS	National Agricultural Statistics Service
NATE	National Association of Tower Erectors
NAWS	National Agricultural Workers Survey
NCCRAHS	National Children's Center for Rural and Agricultural Health and Safety
NCIPC	National Center for Injury Prevention and Control
NEISS	National Electronic Injury Surveillance System
NFPA	National Fire Protection Association
NHTSA	National Highway Traffic Safety Administration
NIH	National Institutes of Health
NIMS	National Incident Management System
NIOSH	National Institute for Occupational Safety and Health
NOIRS	National Occupational Injury Research Symposia
NORA	National Occupational Research Agenda

NRC	National Research Council
NTOF	National Traumatic Occupational Fatality
NTSB	National Transportation Safety Board
OISPA	Occupational Injury Surveillance of Production and Agriculture
OSH Act	Occupational Safety and Health Act
OSHA	Occupational Safety and Health Administration
PASS	Personal Alert Safety System
PWS	proximity warning system
r2p	research to practice
RAND	Research and Development Corporation
RFA	request for application
ROPS	rollover protective structure
SAE	Society of Automotive Engineers
SME	small- to medium-sized enterprise
STFs	slips, trips, and falls
TI	traumatic injury
TISF	Traumatic Injury Surveillance of Farmers
TWU	transportation, warehousing, and utilities
USCG	U.S. Coast Guard
USDA	U.S. Department of Agriculture

Summary

ABSTRACT *Occupational injuries continue to be a significant public health problem in the United States, imposing a substantial human and economic burden. Although rates of both fatal and nonfatal occupational injuries have declined since the passage of the Occupational Safety and Health (OSH) Act in 1970, much remains to be done. In 2006, more than 110 workers died each week as a result of injuries sustained on the job. According to the Bureau of Labor Statistics, in 2006, 3.9 million nonfatal injuries were sustained by U.S. workers in private-sector employment—a number that is widely recognized as a substantial underestimate. The National Institute for Occupational Safety and Health (NIOSH) is the federal agency tasked with conducting research and making recommendations for the prevention of occupational injury and illness.*

The Institute of Medicine convened a committee of experts to review NIOSH's Traumatic Injury (TI) Research Program. The committee evaluated the relevance and impact of the TI Research Program's efforts for the years 1996-2005, reviewed the program's strategic goals for the future, and provided recommendations for program improvement. Using a five-point scoring scale (where 5 is highest), the committee assigned the TI Research Program a score of 4 for both relevance and impact. The committee concluded that research was in priority areas and led to demonstrated effects on some end outcomes or on well-accepted intermediate outcomes. The committee concluded that the TI Research Program's strategic goals for the future were focused on major contributors to occupational injuries and deaths and are sensitive to populations and groups at disproportionate risk. In future iterations of its strategic goals, the TI Research Program should work toward focusing its efforts. The committee developed nine recommendations for program improvement in the areas of strategic planning, coordination and collaboration, workforce development, transfer, and the changing nature of work.

INTRODUCTION

In 2006, 5,840 workers—more than 110 workers each week—died as a result of injuries sustained on the job. These deaths occurred across all industry sectors (BLS, 2008). Nonfatal work-related injuries far outnumber fatalities and are much more difficult to count. According to the Bureau of Labor Statistics (BLS), in 2006, 3.9 million nonfatal injuries were sustained by U.S. workers in private-sector employment (BLS, 2008). More than half of these injuries required workers to transfer to another job, restrict their duties at work, or take time off from work to recuperate. These BLS estimates are widely recognized to underestimate the full extent of the problem. They exclude nonfatal injuries among the 22 percent of the workforce that are not in private-sector employment, and there is also evidence that private-sector injuries are under counted. One population-based study of work injuries (Smith et al., 2005) estimated that counts of injuries resulting in days away from work were 1.4 times higher than BLS workplace-based estimates for the private sector.

NIOSH is a component of the Centers for Disease Control and Prevention (CDC), an agency of the Department of Health and Human Services (HHS). Created in 1970 by the OSH Act along with the Occupational Safety and Health Administration (OSHA) in the Department of Labor (DOL), NIOSH was authorized to

- Conduct research on worker safety and health, including new safety and health problems;
- Develop recommendations for occupational safety and health standards;
- Conduct training and employee education;
- Develop information on safe levels of exposure to toxic materials and harmful physical agents and substances;
- Conduct on-site investigations to determine the toxicity of materials used in workplaces; and
- Fund research by other agencies or private organizations through grants, contracts, and other arrangements.

Congress has clearly distinguished OSHA's functions of regulation and enforcement from NIOSH's research mandate. OSHA's mission is to "assure the safety and health of America's workers by setting and enforcing standards; providing training, outreach, and education; establishing partnerships; and encouraging continual improvement in work-place safety and health."

CHARGE TO THE COMMITTEE

NIOSH has requested that the National Academies, through the National Research Council (NRC) and the Institute of Medicine (IOM), conduct a series of reviews of its research portfolio. This report contains a review and evaluation of the TI Research Program. NIOSH defines work-related traumatic injury as "any damage inflicted to the body by energy transfer during work with a short duration between exposure and health event." Safety research is an interchangeable term in NIOSH publications for traumatic injury research.

The TI Research Program is described by eight goal areas (Box S-1). Within each of the eight goals are two to three subgoals, with the exception of the goal for workplace violence, which does not have any subgoals. Currently, four of the TI Research Program's efforts—agricultural injuries among children, firefighter safety, workplace violence, and workers in Alaska's high-risk industries—are directed by congressional initiatives (NIOSH, 2007, p. 44).

Using an evaluation framework developed by the National Academies Committee to Review the NIOSH Research Programs, the "Framework Committee" (see Appendix A), this committee was to evaluate the relevance and impact (using an integer score from 1-5, with 5 being the highest) of the TI Research Program, as well as its future directions. (See Boxes S-2 and S-3 for more information on the scoring system.) The committee was also encouraged to provide recommendations for program improvement.

The committee was comprised of 10 members and one Framework Committee liaison.[1] The committee had expertise in the areas of occupational health, public health education, medicine, injury prevention and control, epidemiology and biostatistics, labor, industry, program evaluation, ergonomics, and bioengineering. The committee evaluated the TI Research Program for the period 1996-2005, the first decade of the National Occupational Research Agenda (NORA). The information evaluated consisted primarily of materials—organized by traumatic injury goals and subgoals—included in an evidence package provided by NIOSH to the

[1]The Framework Committee liaison is a member of the National Academies Committee to Review the NIOSH Research Programs, which is the committee that developed the framework (see Appendix A of this report) for review of NIOSH research programs.

BOX S-1
Goals and Subgoals of the NIOSH Traumatic Injury Research Program[a]

1. **Reduce injuries and fatalities due to motor vehicles**
    1.1. Reduce occupational injuries and fatalities due to highway motor vehicle crashes
    1.2. Reduce occupational injuries and fatalities due to motor vehicle incidents in highway and street construction work zones
2. **Reduce injuries and fatalities due to falls from elevations**
    2.1. Reduce worker falls from roofs
    2.2. Improve fall-arrest harnesses
    2.3. Reduce worker falls from telecommunications towers
3. **Reduce injuries and fatalities due to workplace violence**
4. **Reduce injuries and fatalities due to machines**
    4.1. Reduce injuries and deaths caused by tractor rollovers by increasing availability and use of effective rollover protective structures
    4.2. Reduce worker injuries and deaths caused by paper balers
    4.3. Reduce injuries and deaths caused by machines through the conduct of fatality investigations and the dissemination of prevention strategies
5. **Reduce acute back injury**
    5.1. Reduce acute injuries caused by patient handling
    5.2. Evaluate interventions used to prevent acute injuries caused by material handling
6. **Reduce injuries and fatalities among workers in Alaska**
    6.1. Reduce injuries and fatalities in commercial fishing
    6.2. Reduce injuries and fatalities in helicopter logging operations
    6.3. Reduce injuries and fatalities in Alaska aviation
7. **Reduce injuries and fatalities to emergency responders**
    7.1. Reduce injuries and fatalities to firefighters
    7.2. Improve protection for ambulance workers in patient compartments
    7.3. Improve protection for emergency workers responding to large-scale disasters and terrorist attacks
8. **Reduce injuries and fatalities to working youth**
    8.1. Influence legislative changes to protect young workers
    8.2. Reduce child agricultural injuries
    8.3. Foster the development and widespread use of safety materials and intervention strategies to protect young workers

[a]The numbering of the goals here is consistent with the numbering of the goals as presented in the evidence package prepared by NIOSH for the committee. The numbering is not a ranking of goals by research priority.

BOX S-2
Scoring Criteria for Relevance

5 = Research is in high-priority subject areas and NIOSH is significantly engaged in appropriate transfer activities for completed research projects or reported research results.
4 = Research is in priority subject areas and NIOSH is engaged in appropriate transfer activities for completed research projects or reported research results.
3 = Research is in high-priority or priority subject areas, but NIOSH is not engaged in appropriate transfer activities; or research focuses on lesser priorities but NIOSH is engaged in appropriate transfer activities.
2 = Research program is focused on lesser priorities and NIOSH is not engaged in or planning some appropriate transfer activities.
1 = Research program is not focused on priorities and NIOSH is not engaged in transfer activities.

BOX S-3
Scoring Criteria for Impact

5 = Research program has made major contribution(s) to worker health and safety on the basis of end outcomes or well-accepted intermediate outcomes.
4 = Research program has made some contributions to end outcomes or well-accepted intermediate outcomes.
3 = Research program activities are ongoing and outputs are produced that are likely to result in improvements in worker health and safety (with explanation of why not rated higher). Well-accepted outcomes have not been recorded.
2 = Research program activities are ongoing and outputs are produced that may result in new knowledge or technology, but only limited application is expected. Well-accepted outcomes have not been recorded.
1 = Research activities and outputs do not result in or are NOT likely to have any application.
NA = Impact cannot be assessed; program not mature enough.

committee. For its assessment of the NIOSH process for targeting new research areas and identifying emerging issues in occupational safety and health, the committee relied primarily on a review of the TI Research Program's strategic goals for the future (a list of these goals was included in the evidence package provided to the committee by NIOSH).

RELEVANCE AND IMPACT OF THE TI RESEARCH PROGRAM

The committee reviewed the work supporting the eight specific goals (including the 19 subgoals) that constitute the TI Research Program. The goals represent a mix of long-standing safety concerns (e.g., agricultural injuries), newer or emerging areas of emphasis for the TI Research Program (e.g., falls from telecommunications towers), and congressionally driven attention to important occupational risks (e.g., the Alaska Field Station [AFS]). Three goals represent specific worker populations identified by location (Alaska), age (youth), or sector (emergency response). The TI Research Program has clearly driven a national sensitivity to some specific safety problems. For example, the committee concludes that TI Research Program attention to workplace violence has highlighted a previously neglected area.

The committee concludes that, for the most part, the TI Research Program goals are appropriate and relevant to the burden of traumatic injury in the workplace. The burden of injury represented by the eight major goal areas is certainly high, although the committee did not attempt to independently assess the burden of injuries in all occupations or worksites in the country as part of its review. Rather, the committee understands the challenges NIOSH faces in prioritizing research with restricted resources and concludes that—given its limited budget—the TI Research Program has made overall appropriate selections of general areas to pursue.

Although the committee concluded that many of the goal areas were high priority (e.g., Alaska, falls from elevations), it identified gaps, particularly within the subgoals (falls from the same elevation; several areas within workplace violence; a narrow focus within machines). The TI Research Program engages in appropriate transfer activity within some, but not all, of the goal areas. In summary, the committee notes impressive work, including transfer, in priority goal areas. The committee assigns a score of 4 for the relevance of the TI Research Program.

The committee commends the TI Research Program for its contributions toward reducing occupational traumatic injuries. The TI Research Program is associated with impact on either intermediate or end outcomes in each major goal. The committee recognizes that external factors—specifically, severely limited resources and inaction on the part of OSHA—can be significant barriers to the TI Research Program's progress in some goal areas. However, the committee notes (1) the lack of demonstrated effect on end outcome data in three goal areas and in some subgoals of the other five goals; (2) the inability to determine what degree of responsibility the TI Research Program bears for the documented improvements in end outcomes or for the intermediate outcomes; and (3) a lack of significant intermediate outcomes for some subgoals. The committee assigns a score of 4 for the impact of the TI Research Program.

TARGETING NEW RESEARCH

The second part of the committee's charge was to perform an assessment of the TI Research Program's effectiveness in targeting new research areas and identifying emerging issues most relevant to future improvements in workplace protection.

TI Research Program project planning takes place at the research division level and at the institute level. At the division level, Division of Safety Research (DSR) staff propose research projects within the context of program drivers, which may include surveillance findings on injury incidence and severity, worker groups with the greatest numbers and risks of death or injury, congressional mandates, stakeholder input, or research needs outlined in the 1998 NORA-TI Research Program team white paper. DSR leadership—with input from staff—rates and ranks new project concepts based on project need, soundness of approach or methods, and expected impact (NIOSH, 2007, p. 43). Staff may then develop research protocols within the approved concept areas. Research protocols are peer-reviewed internally and may also be presented at public meetings for stakeholder input and to assess the interest in and potential impact of the research.

According to DSR leadership, most current TI Research Program research projects are funded through the institute-wide NORA funding competition. DSR "base" funds (annual division or lab allocation) have diminished and are now used primarily for ongoing surveillance and field investigation programs, as well as congressionally mandated projects (NIOSH, 2007, p. 43). There has not been competition for new projects with DSR base funds for the past 3 years.[2]

The TI Research Program has several means by which it receives input from stakeholders on its research programs. For intramural projects, NIOSH frequently holds a public meeting announced in the *Federal Register*. Public meetings may be held to discuss proposed research projects that will develop or evaluate products (versus policies or procedures) that have broad stakeholder vested interest and/or are potentially controversial.[3] The TI Research Program also organizes and hosts periodic National Occupational Injury Research Symposia (NOIRS), which bring together researchers from a broad range of disciplines to discuss research in progress and to form research and prevention partnerships. Workers, advocates, and other nonresearch groups may also attend and have an opportunity to provide input regarding traumatic injury research needs. At the inception of the four TI research programs currently directed by congressional initiatives and mandates, NIOSH held stakeholder meetings to obtain input on possible research directions within these areas.

---

[2]Personal communication from Nancy Stout.

[3]Personal communication from Nancy Stout.

The TI Research Program uses surveillance data on fatal and nonfatal injuries—primarily from the BLS surveillance systems—to identify emerging research needs. The program also has real-time access to data on injuries reported at hospital emergency departments through the National Electronic Injury Surveillance System, as well as data on fatal injuries in selected states through the NIOSH Fatality Assessment and Control Evaluation (FACE) Program, which allow for quicker detection of injury clusters and spikes (NIOSH, 2007, p. 46), as well as sentinel fatalities that identify previously unrecognized hazards.

The NORA process and the relationships with stakeholders that result from it are also an important means of identifying emerging issues. In NORA, a committee specifically addressed traumatic injury research needs; in the current NORA II process, industry sector councils have been formed to address sector-specific research needs. Traumatic injuries are addressed by each of these councils.

The committee is sensitive to the need for the TI Research Program to choose its research activities carefully to make the best use of limited resources. Research project planning should focus on occupational risks that it has specific skills for addressing and which are not currently or better addressed by other federal or nonfederal researchers.

While the committee recognizes that it is important to consider both the severity and the magnitude of injury when setting traumatic injury research priorities, it is concerned that the balance of focus between fatal and nonfatal injuries is either not evident or not optimal. Based on information provided in the evidence package, it is apparent that occupational fatality surveillance data have been an important program driver. Nonfatal occupational injuries far outnumber fatal injuries, and risk factors for fatal and nonfatal injuries are not necessarily the same. Although a focus on fatalities is reasonable in light of limited resources, this leaves a very substantial gap with respect to nonfatal injuries. Additional surveillance and surveillance research are needed to improve the characterization of nonfatal injuries (see Chapter 4 for a discussion).

The TI Research Program's draft strategic goals for the future are to (1) reduce fall injuries in the workplace; (2) reduce occupational injuries and deaths due to motor vehicles; (3) reduce occupational injuries and deaths due to workplace violence; (4) reduce occupational injuries and deaths due to machines and industrial vehicles; and (5) reduce occupational injuries and deaths among high-risk and vulnerable worker groups.[4] Within each of the five strategic goals are three or four subgoals that generally identify types of injuries, worker populations, industries,

[4]The numbering of the goals here is consistent with the numbering of the goals as presented in the evidence package prepared by NIOSH for the committee. Numbering is not a ranking of goals by research priority.

and workplace exposures on which to focus. At the next level are intermediate goals for achieving the objectives of the goals and subgoals.

Based on the expertise of its members, the committee identified areas of research that warrant attention in the future. Some of these are described within the context of the review of the strategic goals, and some are described in a subsequent section. Given limited staffing and budget resources, it is not expected that the TI Research Program will pursue all of the proposed research areas, but rather that it will take them into consideration.

Overall, the committee finds that the TI Research Program's draft strategic goals are focused on major contributors to occupational injuries and deaths and are sensitive to populations and groups at disproportionate risk. Among the draft strategic goals are several intermediate goals for leveraging partnerships that, if carried out, could help the TI Research Program to maximize its impact a great deal.

In future iterations of its strategic goals, the TI Research Program should work toward focusing its efforts (e.g., goals for motor vehicles) on areas that are not researched by other agencies or covered by other agency missions. In general, the committee also feels that future modifications to the goals could include better indication of how proposed interventions and partnerships will be evaluated.

The committee identified areas of research that warrant attention in the future. Some of these are possible gaps in the five strategic goals; others are of a more cross-cutting nature. With regard to Strategic Goal 1: Reduce Falls in the Workplace, the committee was disappointed to see no obvious attention to slips, trips, and falls from the same elevation. The TI Research Program should consider including research on tribology and on risk factors for falls in older workers.

The committee identified gaps in the plans for future work in motor vehicles. The committee urges the TI Research Program to consider taxi driving, short haul trucking, day delivery drivers, parking lot occupational driving, and intrastate driving. These specific areas provide opportunities for synergy (e.g., taxi drivers are at risk for workplace violence injuries and for occupational driving injuries) and fill a gap not addressed by another agency (National Highway Traffic Safety Administration researches long-haul trucking safety, but short-haul trucking is under-researched).

The committee identified gaps in the goal regarding machines and industrial vehicles. The committee urges the TI Research Program to consider augmenting its future goals with research on TI injuries in the landscaping and horticultural industries, which have been identified as one of the most hazardous industries.

The committee urges the TI Research Program to consider augmenting its work on vulnerable workers. The committee identified several groups of particular importance for the future: older workers, immigrant and minority workers, workers aged 18-24 years, and workers with developmental or physical disabilities. In addition, the committee urges the TI Research Program to extend risk factor research on vulnerable workers to include the study of informal and formal workplace policies and workplace norms and alternative work arrangements.

In addition to gaps in research noted within the context of the five strategic goals, the committee identified other priority research areas that the TI Research Program could pursue in the future. These are as follows:

- Organizational culture and adoption of safety measures
- Cost of injuries
- Policy evaluation research
- Small-to-medium sized enterprises (SMEs)
- Surveillance research

RECOMMENDATIONS FOR PROGRAM IMPROVEMENT

After reviewing the evidence package provided by the TI Research Program and evaluating its work in the eight goals, the committee developed a series of general recommendations for program improvement. These fall in the general areas of strategic planning, coordination and collaboration, workforce development, transfer, and the changing nature of work. A summary of the recommendations appears in Box S-4.

Strategic Planning

The current goal areas represent a retrofitting of a decade's worth of work into a structure to be reviewed with the Framework Document in mind.[5] This does not mean that considerable thought was not given by the TI Research Program over the decade to the work being done, but it is clear that some of its efforts occurred outside of and independent of a program-wide coherent planning process.

The committee recognizes that opportunities arise and an agency must be adroit to deal with unexpected events (for example, the work in goal 2 regarding

[5]The TI Research Program presented its portfolio in the evidence package according to these goals in order to conform to the Framework Document (e-mail from N. Stout to K. Stratton, November 30, 2007). Prior to this configuration, the organizing structure for describing the project portfolio were three program areas that reflected (1) the four leading causes of occupational injury death (motor vehicles, machines, violence, and falls), (2) four high-risk industries (construction, transportation, agriculture/forestry/fishing, and services), and (3) five NIOSH/CDC crosscutting priority areas (disparities, economic impact, emergency response, NORA implementation, and emerging issues). Individual projects had relevance to one or more of the 13 elements that comprise those three program areas.

BOX S-4
Summary of Recommendations

Strategic Planning

Continue setting goals that are within the TI Research Program's scope and resources.
Develop an explicit plan for each subgoal.

Coordination and Collaboration

Work with other federal agencies that support injury prevention and control research.
Improve surveillance of nonfatal injuries.
Work collaboratively with OSHA.
Ensure collaboration among NIOSH-funded researchers.

Workforce Development

Increase visibility of traumatic injury research.

Transfer

Evaluate research-to-practice efforts.

The Changing Nature of Work

Research prevention strategies for traumatic injuries in a changing workplace.

falls from telecommunications towers appears to be a response to a newly discovered occupational risk) and that there are points of departure from any planning document. Some work will appear not to fit in well with the rest of the program. This is not unexpected. With some exceptions, as discussed in Chapter 2, the program has worked in areas of public health importance and has documented intermediate outcomes.

The committee concludes that the TI Research Program successes (as defined by the activities, outputs, and outcomes reviewed) have occurred most obviously in goal areas in which there was a focused and intense effort (due to resources, e.g., Alaska) or in which the goal was narrowly or clearly defined and achievements could be documented. Often these areas of focus were due to congressional directives, the increased resources associated with congressional directives, and staff interest and expertise. Some of the successes arose from a focus on a newly emerging concern.

The TI Research Program should be careful in its next stage of planning and priority setting to outline as specifically as possible the scope of the work it plans to accomplish so that its achievements are demonstrably linked to a problem of importance and its research is strategic. The committee urges the TI Research Program to focus on those occupational risks that it has specific skills for addressing and which are not currently or better addressed by other federal or nonfederal researchers. Otherwise, it risks squandering precious resources on activities that could be redundant or that will not necessarily or directly lead to accomplishing the goals of reducing morbidity and mortality from occupational traumatic injuries.

1. **Continue setting goals that are within the TI Research Program's scope and resources.** Given its limited resources, the TI Research Program should continue a research focus and priority setting on goals that are well defined, are based on rigorous surveillance data, and are complementary to work being done by stakeholders, extramural research partners, or other agencies.
2. **Develop an explicit plan for each subgoal.** The TI Research Program should develop an explicit, written plan within each subgoal for progression along the public health framework, including the circumstances under which work in the subgoal should cease. Additional considerations should be the relative balance between risk factor and intervention research.

Coordination and Collaboration

Given that the TI Research Program operates under severely limited resources, it must not only be strategic in selecting its priorities, as discussed previously, but also position itself to benefit from collaborations within the federal government and with academic researchers and state agencies. These collaborations and coordinating activities will help the TI Research Program prioritize its activities in order to complement work elsewhere or to avoid duplication of effort. Coordination and collaboration can be achieved by several means—for example, organizational relationships and research programs. As such, the committee offers several recommendations that are intended to support and encourage some current collaborations, to identify a major new initiative, and to leverage a relationship that has perhaps recently been ignored.

NIOSH is one of several federal agencies with a role in injury prevention and control, so there is obviously some overlap of agency interests, particularly with regard to research and information dissemination. In focusing on occupational traumatic injury, NIOSH should continue to foster and build relationships with other agencies, (e.g., the National Highway Traffic Safety Administration, the National Institutes of Health, the National Center for Injury Control and Prevention, the Consumer Product Safety Commission, the Department of Defense, the Health Resources and Services Administration, and the Agency for Healthcare Research and Quality). The committee encourages continued and appropriate attention to interagency issues in order to assure a unique research portfolio in the TI Research Program and the efficient deployment of scarce resources.

3. **Work with other federal agencies that support injury prevention and control research.** NIOSH and its TI Research Program should work with senior leadership from other federal agencies to outline areas of collaboration and synergy; to identify opportunities to further the science of injury control and prevention; and to reduce the burden of injury across populations, environments, and products.

The TI Research Program has clearly been driven by a focus on fatalities. The committee understands that there are legitimate reasons to focus on fatalities and legitimate debates about aggregate burdens, but the time has come for the TI Research Program to bolster its focus—particularly starting with surveillance—on nonfatal occupational injuries without lessening the excellent work on fatal injuries.

4. **Improve surveillance of nonfatal injuries.** The TI Research Program should develop a plan for improving surveillance of nonfatal injuries, integral to prevention and to strengthening the TI Research Program portfolio development. A comprehensive approach should go beyond use of employer-based data to include nonemployer-based data sources such as hospital data and other medical data systems, the National Health Interview Survey, and the Behavioral Risk Factor Surveillance System. The TI Research Program should involve other relevant federal and state agencies in developing a cohesive interagency effort.

OSHA is a significant external factor to the success of NIOSH research. The committee understands the difficulty NIOSH might experience in this regard. The committee urges perseverance on the part of the TI Research Program in addressing injuries and interventions of interest amenable to regulatory action by OSHA.

5. **Work collaboratively with OSHA.** An agency of particular importance and relevance to NIOSH is OSHA. The TI Research Program, along with NIOSH leadership, should continue to work with OSHA to identify areas of high-priority research that NIOSH should undertake and

to identify NIOSH research findings of particular salience for potential regulatory action by OSHA.

Members of the committee understand the tension between the needs of federal agencies and the desires of academic researchers. Academic researchers are not necessarily interested in the pragmatic questions that federal agencies need addressed. Academic researchers also tend to enjoy a degree of freedom in research settings that might not be compatible with the data or research requirements of an agency with a very narrow and directed mission. Conversely, agency priority setting activities might not allow for sufficient time for nondirected basic research in the academic setting to show relevance to agency needs. Finally, agencies often need results disseminated promptly, which is not always compatible with the timeline of the traditional academic publication process. The intramural program has obvious areas of expertise that should be used to the fullest, while new hires could be used to expand the intramural capability. The extramural research program can "fill in" what the intramural program lacks, as well as provide innovative approaches not currently anticipated or realized.

However, with its limited research budget, the TI Research Program deserves and needs to fully understand the work it is supporting. Communication and collaboration are key to a successful mix of intramural and extramural research. Increased communication can also lead to increased opportunities for appropriate transfer of research findings into practice. The committee urges that this collaboration not be overly directed or programmed. Rather, the committee sees this as an important step in building teams for traumatic injury research. The committee recognizes that there are obvious benefits to increased interactions between researchers and that NIOSH should facilitate such interactions.

6. **Ensure collaboration among NIOSH-funded researchers.** NIOSH should review its practices and take steps to improve the opportunities for intramural and extramural researchers, including state occupational public health programs, to communicate and collaborate without excessively directing extramural research to the detriment of scientific creativity. NIOSH should also further ensure collaboration and coordination among its programs, including the traumatic injury, construction, mining, and agriculture programs.

Workforce Development

Although there are no studies of this issue, the committee feels—through its own experiences and from discussions with fellow traumatic injury researchers—

that it is difficult to attract new students to this very applied field. The TI Research Program needs to develop a means to increase the pipeline of traumatic injury researchers. In order to do so, the committee suggests stressing the technical expertise required to work in traumatic injury, the interdisciplinary nature of successful traumatic injury research, and the possibility of "making a real difference" with traumatic injury research.

7. **Increase the visibility of traumatic injury research.** NIOSH should embark on a program to increase the visibility of traumatic injury research in order to attract new researchers. Absent a significant increase in research funding, the TI Research Program can still attempt to influence the number of Education Resource Centers that have a focus on safety research and can still disseminate information about the quality, impact, and scientific challenges of traumatic injury research, as well as the dynamic changes in the field that go beyond the confines of traditional safety engineering.

Transfer

NIOSH created a research-to-practice (r2p) initiative with six components: prioritize, partner, target, translate, disseminate, and evaluate. This initiative recognizes the role of partners in this collaborative process. The TI Research Program review in Chapter 2 includes several good examples of r2p efforts, most notably the work of the AFS. However, the committee is concerned that the TI Research Program, and perhaps all of NIOSH, is not fully prepared to rigorously and expertly execute an r2p enterprise. In order to improve on this initiative, it is important that the TI Research Program allow its talented staff to focus their efforts, play on their strengths, and collaborate with others to complement their own expertise. Experts in translation should be included in project teams. As the TI Research Program develops better tracking of extramural research projects, translation activities regarding the outcomes of this research can be planned, whether through translation components included in the extramural research or by collaborations with the NIOSH transfer experts. The committee notes that it will be important for the TI Research Program to maintain a balance between basic research, applied research, and transfer activities.

8. **Evaluate research-to-practice efforts.** NIOSH should develop a strategic plan for evaluating its research-to-practice efforts and for building the capacity to carry out and evaluate these efforts. Needed disciplines include behavioral sciences; organizational behavior; intervention ef-

fectiveness research; public health education; dissemination, implementation, and diffusion research; social marketing; and media advocacy.

The Changing Nature of Work

Recognizing the changing demographics of the U.S. workforce, NIOSH has included reducing injuries among high-risk and vulnerable populations among its strategic goals for the future. It is likewise important to address the changing nature of work itself and its interplay with the changing workforce. Work in the United States is changing in significant ways that can be expected to alter the current pattern of and risk factors for work-related traumatic injury. The industrial sectors in which U.S. workers are employed are changing. Most notably, the United States continues to shift generally from a manufacturing to a service and knowledge economy. For example, the greatest growth is projected to occur in home health care, an industry which relies heavily on immigrant and minority labor and in which the work setting is geographically dispersed. Another example is where the "craft" of residential construction is changing to a manufactured or prefabricated industry. In addition, there is likely to be a continued shift in work organization and employment practices including corporate restructuring and downsizing, shifts to leaner, more flexible production methods, and increased reliance on part-time, temporary, and contingent labor. These trends may influence work hours, job demands, benefits, and job security that may in turn adversely impact injury risks and may disproportionately affect vulnerable worker populations. Recent emphasis on the development of new sustainable technologies and green building practices offers important new opportunities for prevention through design in which the health of working people as well as the environment would be taken into account in the design stage of new products and projects.

NIOSH has a cross-sector program on work organization and stress-related disorders, which was also one of the 21 priority areas for research under the initial NORA. This program clearly recognizes the potential impact of the changing organization of work not only on worker health but also worker safety and has developed a research agenda to identify and address these potential risks (NIOSH, 2002). More recently NIOSH has established a program on prevention through design, which is broadly defined as addressing occupational safety and health needs in the design process to prevent or minimize work-related hazards and risks. The committee underscores the importance of TI Research Program collaboration with these other NIOSH program areas as well as the NIOSH Program on Occupational Health Disparities.

9. **Research prevention strategies for traumatic injuries in a changing workplace.** The TI Research Program should consider research on the safety impacts of changes in the nature of work as well as intervention research targeting organization polices and practices and including prevention through design approaches.

CONCLUSION

With a focus on program improvement as outlined in this chapter, the TI Research Program can continue to serve as a leader in the field by identifying its niche in research, collaborating with partners, and sponsoring important high-quality research that contributes to reducing the morbidity and mortality associated with injury in the workplace.

REFERENCES

BLS (Bureau of Labor Statistics). 2008. Injuries, illnesses, and fatalities. http://www.bls.gov/iif/home.htm (accessed July 23, 2008).

NIOSH (National Institute of Occupational Safety and Health). 2002. *The Changing Organization of Work and the Safety and Health of Working People: Knowledge of Gaps and Research Directions.* Washington, DC: HHS.

NIOSH. 2007 (unpublished). *Overview of the TI Research Program.* In the evidence package provided to the Committee to Review the NIOSH TI Research Program. NIOSH.

Smith, G. S., H. M. Wellman, G. S. Sorock, M. Warner, T. K. Courtney, G. S. Pransky, and L. A. Fingerhut. 2005. Injuries at work in the U.S. adult population: Contributions to the total injury burden. *American Journal of Public Health* 95(7):1213-1219.

1

Introduction

The National Institute for Occupational Safety and Health (NIOSH) has requested that the National Academies, through the National Research Council (NRC) and the Institute of Medicine (IOM), conduct a series of reviews of its research portfolio. The charge is discussed in detail in a subsequent section, but in general it is to evaluate the relevance and impact of a NIOSH research program and to provide recommendations for future research. This report contains a review and evaluation of the NIOSH Traumatic Injury (TI) Research Program. NIOSH defines work-related traumatic injury as "any damage inflicted to the body by energy transfer during work with a short duration between exposure and the health event" (NIOSH, 2007a, p. 3). Safety research is an interchangeable term in NIOSH publications for traumatic injury research. Traumatic injury is distinguished from psychological trauma and from musculoskeletal injuries caused by repetitive trauma.

This chapter begins with an overview of the impact of occupational injury, and the mission and organization of NIOSH and the TI Research Program, including the relationship of NIOSH to other relevant agencies. The chapter progresses to discuss the TI Research Program goals. A discussion of the charge to the committee and the framework under which it completed this review concludes the chapter.

IMPACT OF OCCUPATIONAL INJURIES

Occupational injuries continue to be a significant public health problem in the United States, imposing a substantial human and economic burden. Although rates of both fatal and nonfatal occupational injuries have declined since the passage of the OSH (Occupational Safety and Health) Act in 1970, much remains to be done. In 2006, 5,840 workers—more than 110 workers each week—died as a result of injuries sustained on the job. These deaths occurred across all industry sectors (BLS, 2007a). Nonfatal work-related injuries far outnumber fatalities and are much more difficult to count. According to the Bureau of Labor Statistics (BLS), in 2006, 3.9 million nonfatal injuries were sustained by U.S. workers in private-sector employment (BLS, 2007b). The injury rate based on this number was 4.2 per 100 full-time workers. More than half of these injuries required workers to transfer to another job, restrict their duties at work, or take time off from work to recuperate. These BLS estimates are widely recognized to underestimate the full extent of the problem. They exclude nonfatal injuries among the 22 percent of the workforce that are not in private-sector employment, and there is also evidence that private-sector injuries are undercounted. One population-based study of work injuries (Smith et al., 2005) estimated that counts of injuries resulting in days away from work were 1.4 times higher than BLS workplace-based estimates for the private sector.

Direct workers' compensation costs for occupational injuries and illnesses that occurred in the United States in 2005 were estimated to be $88.4 billion, a 2.3 percent increase over the estimate for 2004 (Sengupta et al., 2007). This number reflects only a portion of the economic costs of workplace injuries and illnesses. Researchers have found that many injuries never enter the workers' compensation system and that workers' compensation benefits cover only a fraction of lost earnings (Boden and Ozonoff, 2008; Rosenman et al., 2006; Silverstein et al., 1997). A recent study shows that in five states studied, only between 29 and 46 percent of lost earnings were replaced (Reville et al., 2001). Finally there may be substantial additional costs related to the disruption of work and the hiring and training of new workers (Leigh et al., 2000).

In recent years, researchers have also begun to conceptualize and demonstrate empirically the social and economic consequences of occupational injuries for the lives of individual workers and their families. Long-term financial hardship, changes in injured worker domestic function as well as ability to perform activities of daily living, and shifts in family dynamics associated with loss of income and disability are among the outcomes that have been noted (Dembe, 2001; Hensler et al., 1991; Morse et al., 1998; Pransky et al., 2000; Strunin and Boden, 2004).

MISSION AND ORGANIZATION OF NIOSH AND THE TI RESEARCH PROGRAM

NIOSH is a component of the Centers for Disease Control and Prevention (CDC), an agency of the Department of Health and Human Services (HHS). Created in 1970 by the OSH Act along with the Occupational Safety and Health Administration (OSHA) in the Department of Labor (DOL), NIOSH was authorized to

- Conduct research on worker safety and health, including new safety and health problems;
- Develop recommendations for occupational safety and health standards;
- Conduct training and employee education;
- Develop information on safe levels of exposure to toxic materials and harmful physical agents and substances;
- Conduct onsite investigations to determine the toxicity of materials used in workplaces; and
- Fund research by other agencies or private organizations through grants, contracts, and other arrangements.

Congress has clearly distinguished OSHA's functions of regulation and enforcement from NIOSH's primary research mandate. OSHA's mission is to "assure the safety and health of America's workers by setting and enforcing standards; providing training, outreach, and education; establishing partnerships; and encouraging continual improvement in workplace safety and health" (DOL, 2008). Although both agencies have a mandate regarding training and education, and NIOSH is charged to recommend standards to OSHA,[1] in general NIOSH is a research agency and OSHA a regulatory agency. This distinction in mandate allows NIOSH research to proceed without pressures regarding the financial or regulatory implications.

NIOSH offices are located across the country, specifically in Washington, D.C.; Pittsburgh, Pennsylvania; Spokane, Washington; Cincinnati, Ohio; Atlanta, Georgia; and Morgantown, West Virginia (see Figure 1-1). The primary organizational units are divisions and laboratories. The programmatic organization, however, recently underwent changes and now reflects a matrix-management approach[2] (see Table 1-1). The organization currently consists of sector programs (e.g., construction, mining) and cross-sector programs (e.g., traumatic injury, respiratory diseases).[3] Most of the work in traumatic injury occurs through the Division of Safety Research (DSR), which conducts or oversees most of the research that is represented

[1]NIOSH has issued few recommendations for standards, known as Criteria Documents, in recent years, presumably because it is a resource-intensive activity toward which OSHA paid demonstrably little attention.

[2]A discussion of the pros and cons of matrix management can be found in Chapter 2.

[3]At the time of the development of its evidence package and as depicted in Table 1-1, NIOSH described its organizational structure as also including coordinated emphasis areas (e.g., economics, occupational health disparities). These areas now are included under cross-sector programs.

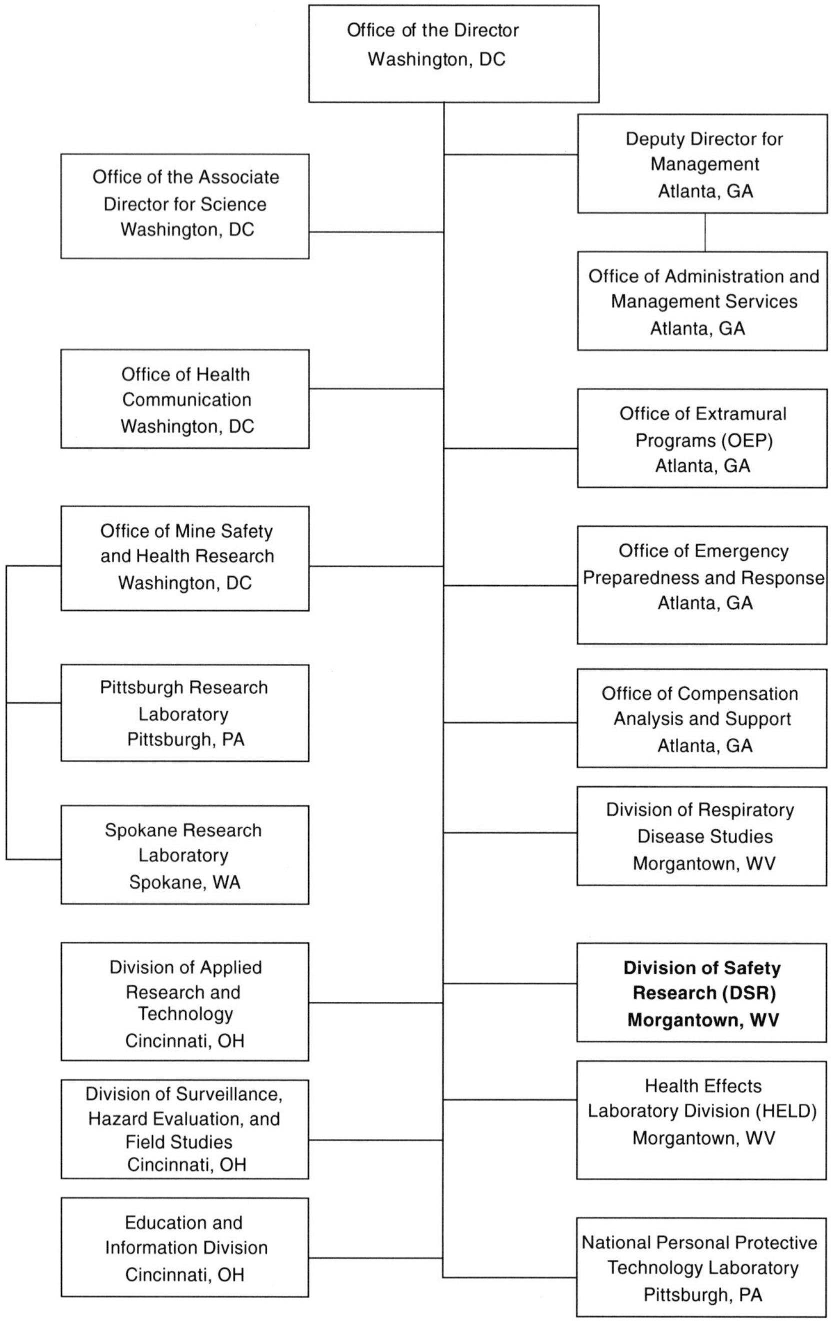

FIGURE 1-1 Organizational components of NIOSH.
SOURCE: NIOSH, 2007c, p. 15

TABLE 1-1 Matrix Management Components of the NIOSH Program Portfolio

NORA Sector Programs	NIOSH Cross-Sector Programs	NIOSH Coordinated Emphasis Areas[a]
Agriculture, Forestry and Fishing	Authoritative Recommendations	Economics
Construction	Cancer, Reproductive and Cardiovascular	Engineering Controls
Healthcare and Social Assistance	Communications and Information Dissemination	Exposure Assessment
Manufacturing	Emergency Preparedness/Response	Occupational Health Disparities
Mining	Global Collaborations	Small Business Assistance and Outreach
Services	Health Hazard Evaluation (HHE)	Surveillance
Transportation, Warehousing and Utilities	Hearing Loss Prevention	WorkLife Initiative
Wholesale and Retail Trade	Immune and Dermal	
	Musculoskeletal Disorders	
	Personal Protective Technology	
	Radiation Dose Reconstruction	
	Respiratory Diseases	
	Training Grants	
	Traumatic Injury	
	Work Organization and Stress-Related Disorders	

[a]See Footnote 3.
SOURCE: NIOSH, 2007i, p. 16.

by the TI Research Program. The TI Research Program consists of intramural research and related activities and extramural research conducted often, but not exclusively, through universities. The TI Research Program also supports state-based surveillance programs. These funds go to state health agencies or state labor departments. Research on traumatic injury from the mining industry, however, is located in the Mining Safety and Health Program and was not reviewed by this committee.

NIOSH is not the only federal agency to sponsor research on traumatic injuries. A companion unit of the CDC, the National Center for Injury Prevention and Control (NCIPC) addresses nonoccupational injuries that are unintentional and those that are due to violence (intentional). NCIPC is located in the CDC headquarters in Atlanta, Georgia. Other federally sponsored surveillance and research relevant to the TI Research Program is carried out by the BLS (DOL), which gathers important data on the incidence of fatal and nonfatal work-related injuries; the National Highway Traffic Safety Administration (Department of Transportation), which funds research on transportation-related occupational traumatic injuries; the Department of Defense; and the National Institute of Justice (Department of Justice), which funds research on workplace violence.

Non-federally sponsored research in traumatic injuries occurs throughout the country, generally but not exclusively in universities. Funding comes from foundations (e.g., Traffic Injury Research Foundation, the National Safety Council), industry (e.g., Ford Motor Company), insurance companies (e.g., Liberty Mutual Research Institute for Safety, the Insurance Institute for Highway Safety), labor unions (e.g., Laborer's International Union of North America), and state governments, although some of these state funds come to the state health and labor departments from the federal government.

The National Occupational Research Agenda

On its twenty-fifth anniversary, NIOSH unveiled the National Occupational Research Agenda (NORA) as an "effort to guide and coordinate research nationally—not only for NIOSH, but for the entire occupational safety and health community" (NIOSH, 1998). Diverse parties collaborated to identify the most critical issues in workplace safety and health. Partners then worked together to develop goals and objectives for addressing those needs. Participation in NORA was broad and included stakeholders from many areas such as universities, large and small businesses, professional societies, government agencies, and worker organizations. NORA identified 21 priority areas, one of which was traumatic injury. Partnership teams were created to develop research agendas for each priority area.

The NORA-TI team released its report in 1998. The report's goal was to present a broad framework of the objectives and research needed to begin filling the gaps in knowledge and furthering progress toward safer workplaces and practices. NORA-TI issued three findings (NIOSH, 1998):

1. Priorities for research and prevention efforts should be based on relevant data. The criteria for setting priorities should include the magnitude of the problem (frequency of injury and size of affected workforce), the risk to workers (rates of injury), injury severity, and amenability to prevention, including cost-effectiveness and likelihood of adoption of prevention strategies by industry.
2. Collaboration among multiple professional disciplines, industries, agencies, and groups is essential to developing research priorities and implementing prevention strategies aimed at reducing occupational injuries.
3. New methods of data collection, data analysis, and information dissemination need to be developed to advance the field of occupational injury research.

Using a five-component public health approach as an organizing framework,[4] the NORA-TI report identified 45 specific research objectives intended to advance occupational injury prevention. Funding for NORA-related projects derived from congressional appropriations earmarked for NORA projects and from other federal agencies. Research teams within NIOSH competed for NORA funds.

NORA recently entered its second decade (often denoted as NORA II) with a sector-based structure, intended to better move research to practice in workplaces. The national agenda will be developed and implemented through eight councils,[5] which will oversee aggregates derived from the 20 North American Industry Classification System sectors. NORA II also includes a Cross-Sector Council, which will set priorities that affect multiple sectors and groups of workers.

Overview of NIOSH and TI Research Program Resources

In fiscal year (FY) 1997 the TI Research Program had 71.62 full-time equivalent (FTE) staff devoted to the eight major TI research goals (to be discussed later). Intramural TI research funding—including the cost of intramural FTEs—was $6,543,305 and extramural research funding was $2,156,591, for a total TI research budget of $8,699,896. (The total NIOSH budget that year was $173 million.) In FY2005, TI Research Program FTEs numbered 87.71; its intramural

[4]Surveillance; analytic risk factor research; prevention and control research; communication, dissemination, and technology transfer; and evaluation.

[5]Agriculture, forestry, and fishing; construction; health care and social assistance; manufacturing; mining; services; transportation, warehousing, and utilities; and wholesale and retail trade.

budget was $11,836,905 and its extramural research budget was $5,369,171, for a total of $17,206,076. (The total NIOSH budget that year was $286 million.) See Figure 1-2 for TI Research Program funding from 1997 to 2005. In FY2005, the total CDC budget was more than $4 billion, the National Institutes of Health budget was over $28 billion, and the total HHS budget was more than $570 billion (HHS, 2005). After adjusting for inflation and earmarks for NORA priorities, the NIOSH budget appears to have stayed relatively stable (see Table 1-2 for a yearly breakdown of TI research resource allocation). Chapter 2 includes a description of TI resources by strategic goal. The committee focuses its review beginning with the establishment of NORA and does not review the entire funding history of NIOSH. Resources attributed to the TI Research Program are those reported by NIOSH in its evidence package and include relevant resources attributed to other NIOSH programs (e.g., construction; agriculture, forestry, and fishing).

NIOSH TI RESEARCH PROGRAM GOALS

NIOSH asked the committee to review its TI Research Program in eight goal areas (Box 1-1).[6] Within each of the eight goals are also two to three subgoals, with the exception of the goal for workplace violence, which does not have any subgoals. The goals outlined in the evidence package provided to the committee represent the four leading causes of occupational injury and death (motor vehicles, machines, violence, and falls); three additional areas shaped by congressional interest (workers in Alaska, emergency responders, and working youth); and a long-standing program in back injury. DSR and the TI Research Program conduct work in areas not represented in this review (e.g., electrocutions and burns), because they are not a substantial part of the portfolio.

According to the evidence package, programmatic focus on motor vehicles, falls, workplace violence, machines, back injury, and workers in Alaska emerged principally from surveillance data on fatal and nonfatal traumatic injuries (NIOSH, 2007a, p. 2). Sources of fatality data include the BLS Census of Fatal Occupational Injuries, as well as NIOSH's National Traumatic Occupational Fatality Surveillance

[6]The TI Research Program presented their portfolio in the evidence package according to these goals in order to conform to the Framework Document (e-mail from N. Stout to K. Stratton, November 30, 2007). Prior to this configuration, the organizing structure for describing the project portfolio were three program areas that reflected (1) the four leading causes of occupational injury death (motor vehicles, machines, violence, and falls), (2) four high-risk industries (construction, transportation, agriculture/forestry/fishing, and services), and (3) five NIOSH/CDC crosscutting priority areas (disparities, economic impact, emergency response, NORA implementation, and emerging issues). Individual projects had relevance to one or more of the 13 elements that comprise those three program areas.

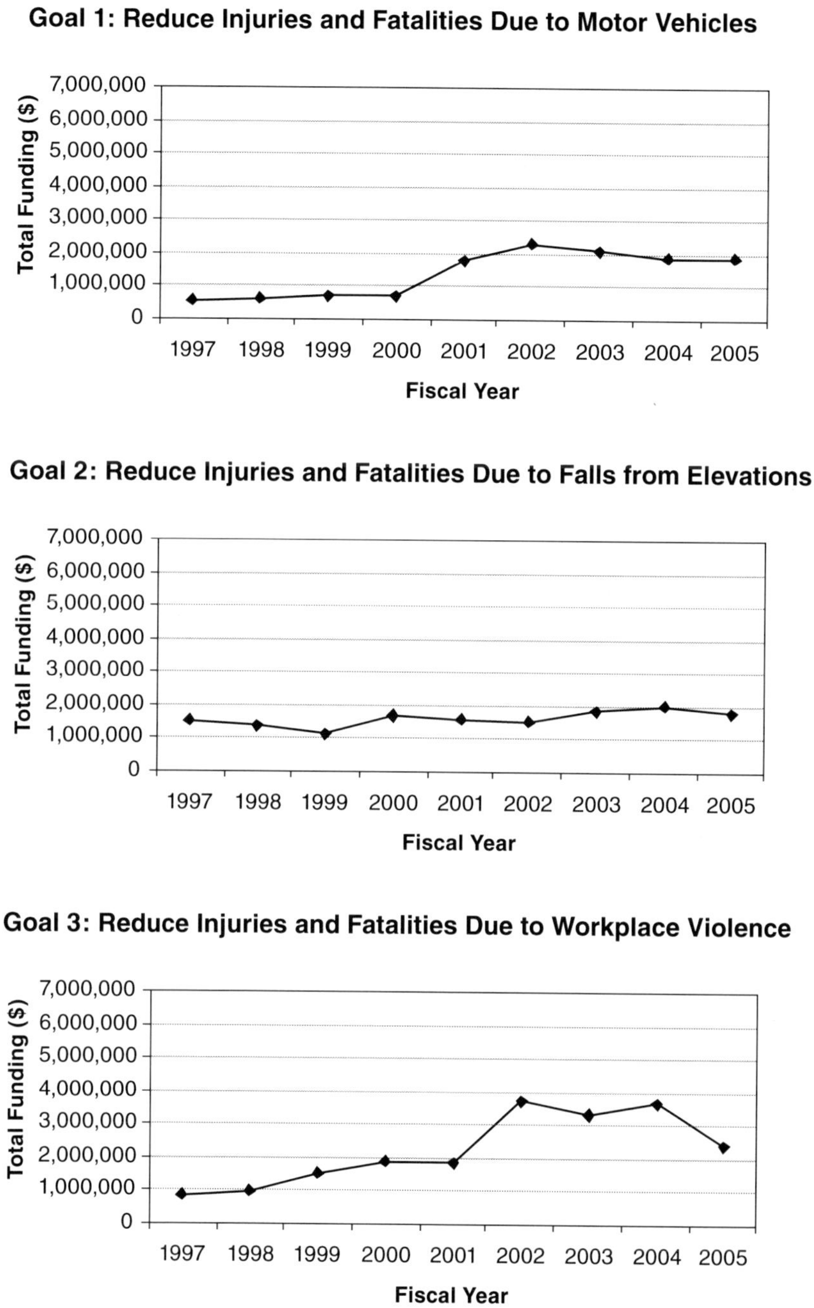

FIGURE 1-2 Combined intramural and extramural funding for individual goals and for all goals combined.

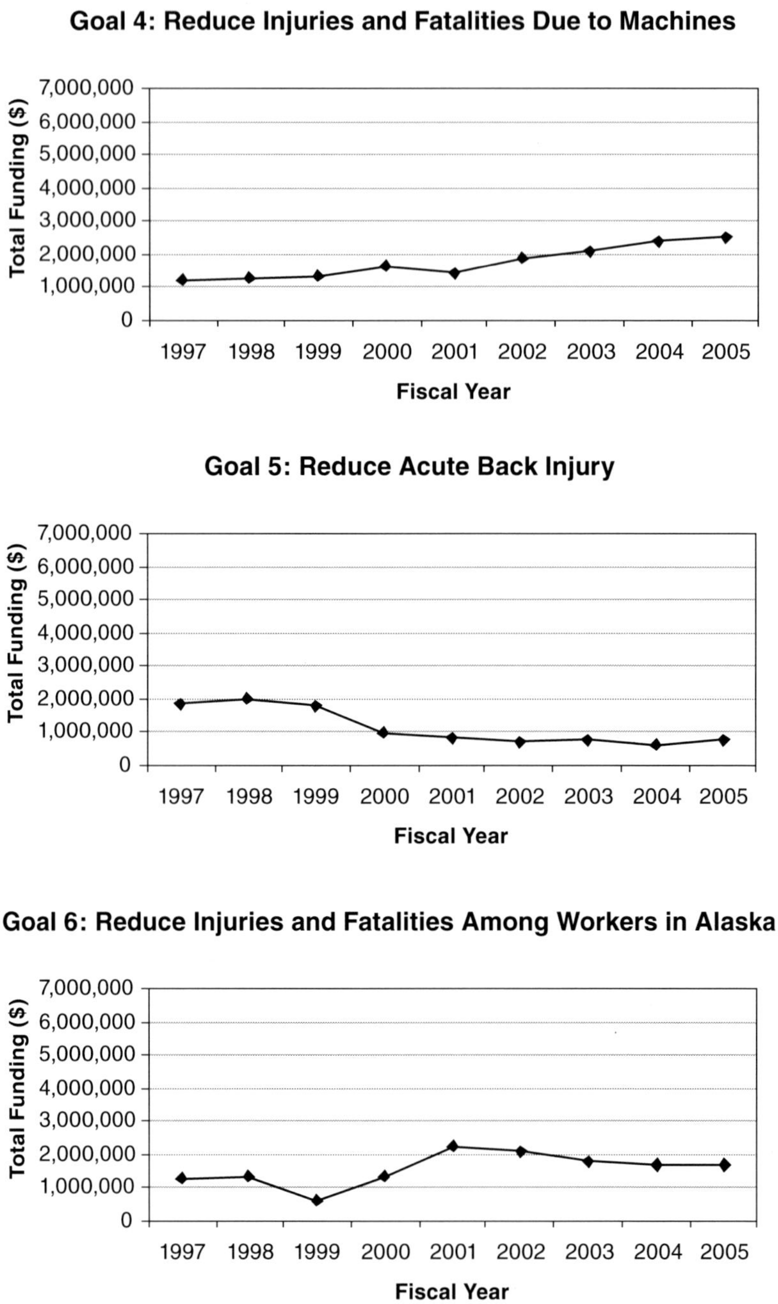

FIGURE 1-2 (continued)

Goal 7: Reduce Injuries and Fatalities to Emergency Responders

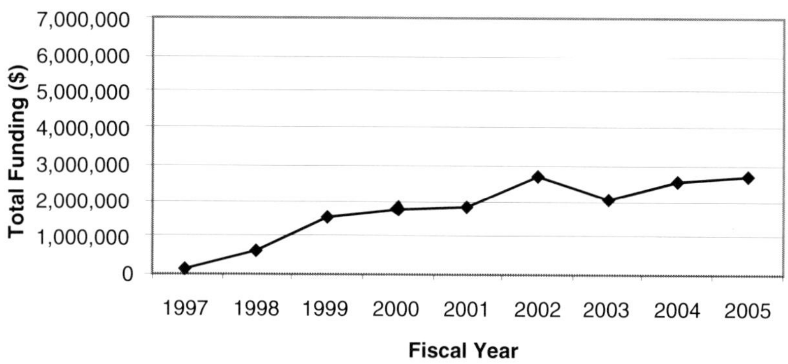

Goal 8: Reduce Injuries and Fatalities to Working Youth

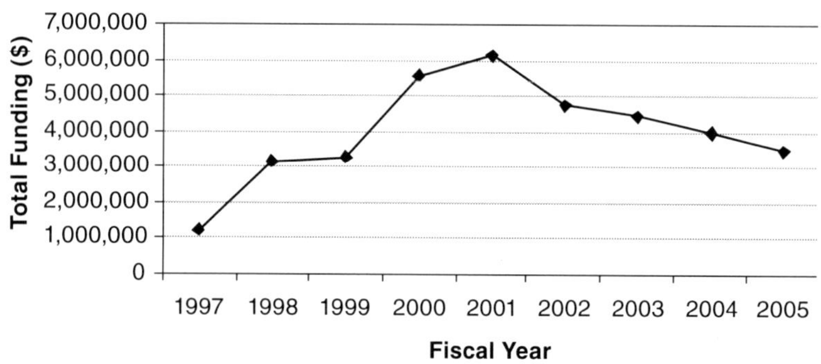

FIGURE 1-2 (continued)

System. Sources of nonfatal injury data include the BLS Annual Survey of Occupational Injuries and Illnesses and the National Electronic Injury Surveillance System managed by the Consumer Product Safety Commission (NIOSH, 2007b, p. 6). In addition to surveillance data, programmatic focus evolved from findings from the Fatality Assessment and Control Evaluation Program, pilot or exploratory studies, and strategic planning work groups or action teams[7] (see Box 1-2 for a snapshot of the evidence of the burden of traumatic injury within the eight programmatic goals). A more detailed picture is presented in Chapter 2.

[7]E-mail from N. Stout to K. Stratton, November 30, 2007.

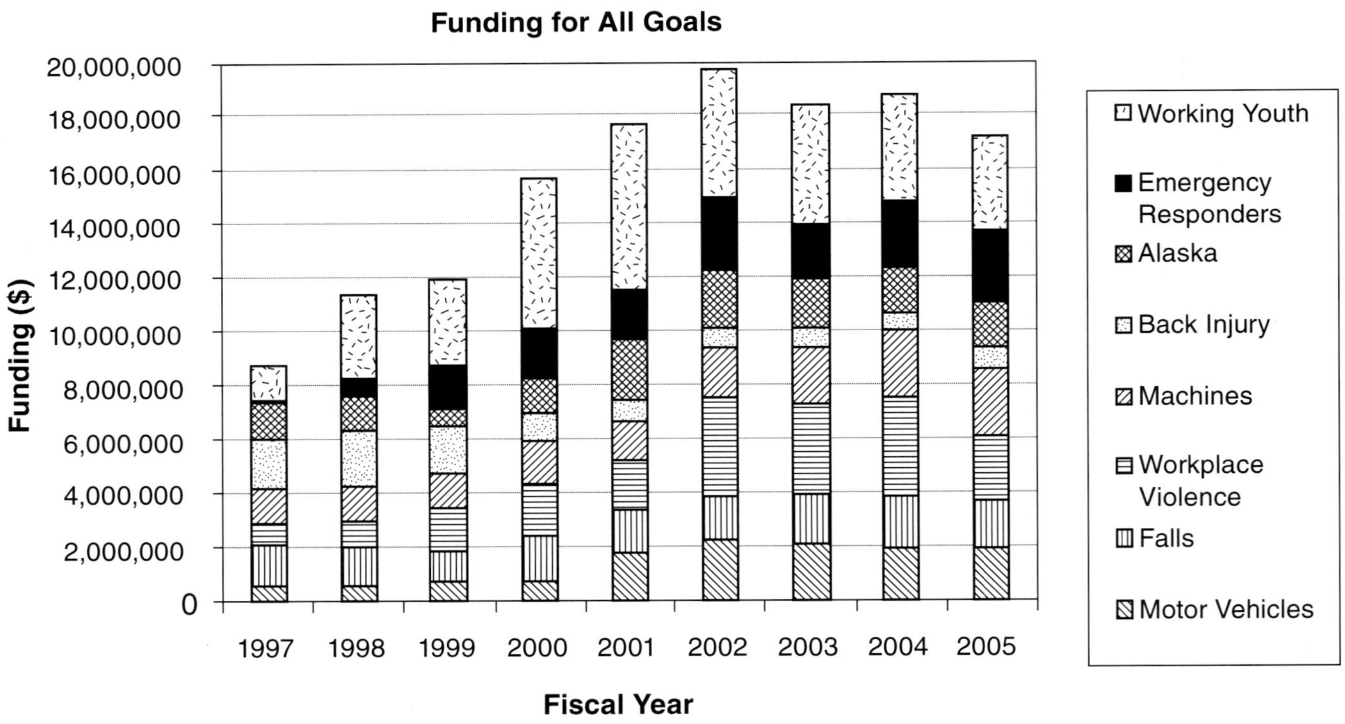

FIGURE 1-2 (continued)

Currently, four of the TI Research Program's efforts—agricultural injuries among children, firefighter safety, workplace violence, and workers in Alaska's high-risk industries—are directed by congressional initiatives (NIOSH, 2007c, p. 44). Specific congressional language follows:

1. Alaska (FY1990): "The committee urges NIOSH to send a team to Alaska to engage in research, prevention and outreach activities to reduce these unacceptably high levels of preventable occupational injuries."
2. Child agriculture (FY1997): "for NIOSH to implement a national plan for childhood agricultural injury prevention initiatives, including research, public education, professional training, community and family services, and policy enforcement."
3. Firefighters (FY1998): "The conference agreement provides increases above the 1997 level for occupational safety and health for the following activities: intramural research at the Morgantown, West Virginia facility; the fire fighter safety initiative; and the national occupational research agenda."
4. Violence (FY2002): "for NIOSH to develop an intramural and extramural prevention research program that will target all aspects of workplace violence and to coordinate its efforts with the Departments of Justice and Labor."

STUDY CHARGE AND EVALUATION COMMITTEE

In September 2004, NIOSH contracted with the National Academies to perform a series of evaluations of its research programs. The IOM and the NRC Division on Earth and Life Sciences were assigned to the task and will convene evaluation committees (ECs) to perform evaluations of as many as 15 NIOSH research programs over a period of 5 years. NIOSH intends to use the findings from these evaluations to increase the impact of its research aimed at reducing workplace illnesses and injuries and improving occupational safety and health.

The TI Research Program was one of the NIOSH programs planned for review and, in March 2007, the IOM convened the Committee to Evaluate the NIOSH TI Research Program. The committee's charge was to perform the following tasks:

1. Assessment of the program's contribution, through occupational safety and health research, to reductions in workplace hazardous exposures, illnesses, or injuries through

TABLE 1-2 TI Research Program Budget by Research Goal

Goal	FY1997	FY1998	FY1999	FY2000
1. Reduce injuries and fatalities due to motor vehicles				
FTE	5.44	4.27	7.99	4.93
Intramural	$318,528	$364,968	$497,628	$669,190
Extramural	$253,962	$244,912	$229,942	$69,679
Total	**$572,490**	**$609,880**	**$727,570**	**$738,869**
2. Reduce injuries and fatalities due to falls from elevations				
FTE	8.31	8.03	13.42	15.43
Intramural	$1,276,334	$922,468	$907,054	$1,459,034
Extramural	$253,962	$463,061	$229,942	$225,879
Total	**$1,530,296**	**$1,385,529**	**$1,136,996**	**$1,684,913**
3. Reduce injuries and fatalities due to workplace violence				
FTE	8.02	4.51	7.28	6.98
Intramural	$602,689	$482,026	$509,517	$653,934
Extramural	$253,962	$511,907	$1,043,966	$1,229,205
Total	**$856,651**	**$993,933**	**$1,553,483**	**$1,883,139**
4. Reduce injuries and fatalities due to machines				
FTE	11.21	9.63	12.77	14.43
Intramural	$993,113	$1,044,805	$941,902	$1,283,376
Extramural	$253,962	$235,037	$365,826	$365,942
Total	**$1,247,075**	**$1,279,842**	**$1,307,728**	**$1,649,318**
5. Reduce acute back injury				
FTE	13.54	15.74	14.72	6.00
Intramural	$1,196,364	$1,356,204	$1,150,931	$597,615
Extramural	$620,369	$654,095	$628,253	$392,677
Total	**$1,816,733**	**$2,010,299**	**$1,779,184**	**$990,292**
6. Reduce injuries and fatalities among workers in Alaska				
FTE	10.58	9.74	6.56	6.58
Intramural	$1,023,592	$1,109,425	$405,931	$757,099
Extramural	$253,962	$228,024	$229,942	$575,870
Total	**$1,277,554**	**$1,337,449**	**$635,873**	**$1,332,969**
7. Reduce injuries and fatalities to emergency responders				
FTE	2.60	2.45	15.03	13.77
Intramural	$154,230	$187,711	$1,074,781	$1,284,491
Extramural	$12,450	$390,299	$480,565	$494,238
Total	**$166,680**	**$578,010**	**$1,555,346**	**$1,778,729**

FY2001	FY2002	FY2003	FY2004	FY2005	Total
7.90	9.95	10.80	14.18	14.79	80.25
$855,111	$1,318,354	$1,229,997	$1,606,772	$1,907,736	$8,768,284
$927,921	$984,331	$877,727	$299,919	$1,250	$3,889,643
$1,783,032	**$2,302,685**	**$2,107,724**	**$1,906,691**	**$1,908,986**	**$12,657,927**
13.56	9.39	14.03	10.79	15.25	108.21
$1,315,445	$1,237,968	$1,568,049	$1,693,086	$1,769,344	$12,148,782
$289,011	$278,004	$285,319	$275,532	$250	$2,300,960
$1,604,456	**$1,515,972**	**$1,853,368**	**$1,968,618**	**$1,769,594**	**$14,449,742**
6.66	7.75	7.56	9.48	8.61	66.85
$839,148	$1,490,096	$1,063,828	$1,273,771	$1,102,187	$8,017,196
$997,606	$2,224,849	$2,233,579	$2,430,217	$1,312,103	$12,237,394
$1,836,754	**$3,714,945**	**$3,297,407**	**$3,703,988**	**$2,414,290**	**$20,254,590**
12.81	14.78	10.48	13.13	12.09	111.33
$1,160,389	$1,593,009	$1,169,991	$1,514,190	$1,479,104	$11,179,879
$289,011	$278,004	$933,245	$890,920	$1,033,342	$4,645,289
$1,449,400	**$1,871,013**	**$2,103,236**	**$2,405,110**	**$2,512,446**	**$15,825,168**
5.16	4.46	5.05	4.83	3.60	73.10
$615,257	$625,395	$647,926	$618,559	$521,036	$7,329,287
$194,526	$99,990	$107,222	$0	$231,250	$2,928,382
$809,783	**$725,385**	**$755,148**	**$618,559**	**$752,286**	**$10,257,669**
12.04	12.80	11.09	11.49	11.81	92.69
$1,493,700	$1,546,891	$1,494,324	$1,400,061	$1,701,203	$10,932,226
$729,000	$568,419	$296,425	$275,532	$0	$3,157,174
$2,222,700	**$2,115,310**	**$1,790,749**	**$1,675,593**	**$1,701,203**	**$14,089,400**
12.32	12.49	18.49	15.02	13.84	106.01
$1,521,576	$2,305,996	$1,932,990	$1,780,148	$1,848,856	$12,090,779
$300,972	$376,491	$101,250	$726,364	$806,938	$3,689,567
$1,822,548	**$2,682,487**	**$2,034,240**	**$2,506,512**	**$2,655,794**	**$15,780,346**

TABLE 1-2 (continued)

Goal	FY1997	FY1998	FY1999	FY2000
8. Reduce injuries and fatalities to working youth				
FTE	11.92	9.71	13.28	12.63
Intramural	$978,455	$1,586,394	$1,778,041	$1,697,255
Extramural	$253,962	$1,559,694	$1,470,082	$3,896,916
Total	**$1,232,417**	**$3,146,088**	**$3,248,123**	**$5,594,171**
Total FTEs for all research goals	71.62	64.08	91.05	80.75
Total intramural budget	$6,543,305	$7,054,001	$7,265,785	$8,401,994
Total extramural budget	$2,156,591	$4,287,029	$4,678,518	$7,250,406
Total budget for all research goals	$8,699,896	$11,341,030	$11,944,303	$15,652,400

BOX 1-1
Goals and Subgoals of the NIOSH Traumatic Injury Research Program[a]

1. **Reduce injuries and fatalities due to motor vehicles (2000[b])**
   1.1. Reduce occupational injuries and fatalities due to highway motor vehicle crashes
   1.2. Reduce occupational injuries and fatalities due to motor vehicle incidents in highway and street construction work zones
2. **Reduce injuries and fatalities due to falls from elevations (1993[b])**
   2.1. Reduce worker falls from roofs
   2.2. Improve fall-arrest harnesses
   2.3. Reduce worker falls from telecommunications towers
3. **Reduce injuries and fatalities due to workplace violence[c] (1991[b])**
4. **Reduce injuries and fatalities due to machines (1990[b])**
   4.1. Reduce injuries and deaths caused by tractor rollovers by increasing availability and use of effective rollover protective structures
   4.2. Reduce worker injuries and deaths caused by paper balers
   4.3. Reduce injuries and deaths caused by machines through the conduct of fatality investigations and the dissemination of prevention strategies
5. **Reduce acute back injury (1985[b])**
   5.1. Reduce acute injuries caused by patient handling
   5.2. Evaluate interventions used to prevent acute injuries caused by material handling

FY2001	FY2002	FY2003	FY2004	FY2005	Total
12.71	8.16	8.14	8.29	7.72	92.56
$1,839,280	$1,561,840	$1,607,812	$1,597,210	$1,507,439	$14,153,726
$4,297,638	$3,192,059	$2,832,167	$2,373,118	$1,984,038	$21,859,674
$6,136,918	**$4,753,899**	**$4,439,979**	**$3,970,328**	**$3,491,477**	**$36,013,400**
83.16	79.78	85.64	87.21	87.71	731
$9,639,906	$11,679,549	$10,714,917	$11,483,797	$11,836,905	$84,620,159
$8,025,685	$8,002,147	$7,666,934	$7,271,602	$5,369,171	$54,708,083
$17,665,591	$19,681,696	$18,381,851	$18,755,399	$17,206,076	$139,328,242

6. **Reduce injuries and fatalities among workers in Alaska[c] (1991[b])**
   - 6.1. Reduce injuries and fatalities in commercial fishing
   - 6.2. Reduce injuries and fatalities in helicopter logging operations
   - 6.3. Reduce injuries and fatalities in Alaska aviation
7. **Reduce injuries and fatalities to emergency responders (1997[b])**
   - 7.1. Reduce injuries and fatalities to firefighters[c]
   - 7.2. Improve protection for ambulance workers in patient compartments
   - 7.3. Improve protection for emergency workers responding to large-scale disasters and terrorist attacks
8. **Reduce injuries and fatalities to working youth (1994[b])**
   - 8.1. Influence legislative changes to protect young workers
   - 8.2. Reduce child agricultural injuries[c]
   - 8.3. Foster the development and widespread use of safety materials and intervention strategies to protect young workers

[a]The numbering of the goals here is consistent with the numbering of the goals as presented in the evidence package prepared by NIOSH for the committee. The numbering is not a ranking of goals by research priority.
[b]The date indicates the year this work was established as a strategic research program area. The committee's review covers only the period of 1996-2005.
[c]Goals/subgoals established or otherwise shaped by congressional initiative.

BOX 1-2
Selected Data on Burden of Traumatic Injury

Motor Vehicles

- Transportation incidents are the leading cause of occupational fatalities in the United States.
- Of the 5,734 fatal occupational injuries that occurred in 2005, 2,493 (43.4 percent) were due to transportation incidents and, of those, more than half (57.6 percent) were highway incidents (BLS, 2007a).
- BLS data indicate that for the period 1995-2002, three out of four worker deaths on road construction sites involved a worker being struck by a vehicle or mobile equipment (Pegula, 2004).

Falls from Elevations

- Falls disable 313,000 American workers each year (NIOSH, 2007d, p. 69) and for the period 1992-2006 resulted in an average of more than 700 occupational fatalities each year (BLS, 2007a).
- Falls from roofs are a common cause of nonfatal injury in the construction industry and are the leading cause of fall fatalities (Hsiao, 2007; NIOSH, 2007d, p. 70).
- Fatality rates due to falls among tower erectors have been estimated to be between 10 and 100 times the average fatality rate across all industries (NIOSH, 2007d, p. 78).

Workplace Violence

- Workplace homicide has been between the second and fourth leading cause of occupational fatalities for the last 15 years (Amandus, 2007).
- In 2005, there were 14,560 nonfatal occupational injuries in private industry involving lost work days due to assaults and violent acts by persons (BLS, 2007a).

Machines

- Based on BLS data through 2005, approximately 770 workers are fatally injured by a machine, plant or industrial powered vehicle, or tractor each year in the United States. Tractors were the leading cause of such deaths (NIOSH, 2007e, p. 97).
- Of the 49 deaths that occurred between 1986 and 2002 due to operation of recycling industry balers, 59 percent involved balers that were baling paper and cardboard (NIOSH, 2007e, p. 104).

Back Injury

- Back injury is one of the most common work-related injuries, accounting for one in five of all private-sector work injuries and illnesses requiring days away from work in 2005 (BLS, 2007a).

- Nurses and other patient caregivers are at increased risk for back injury because of the bending and lifting requirements of patient handling. Among nurses, back pain has been found to be a major factor in choice to leave the profession (NIOSH, 2007f, p. 114).

Alaska

- For the period 1980-1989, the average annual traumatic occupational fatality rate in Alaska was close to 35 deaths per 100,000 workers, almost five times the U.S. average of 7 deaths per 100,000 in the same period (NIOSH, 2007g, p. 126).
- Factors contributing to Alaska's high occupational fatality rate include high-risk commercial fishing and helicopter-logging industries, and aviation crashes associated with the state's dependence upon air transportation, unique terrain and weather challenges, and incomplete radar coverage (NIOSH, 2007g, pp. 126-145).

Emergency Responders

- Each year more than 50 firefighters die in the line of duty from traumatic injury, and another 95,000 sustain a nonfatal injury (NIOSH, 2007h, p. 148).
- Ambulance workers are at risk for crash-related injuries because the use of lap belts in patient compartments does not allow workers the mobility needed to access patients. In addition, equipment in ambulances is often not secured and poses a second hazard to ambulance workers should a crash occur (NIOSH, 2007h, p. 157).
- Ensuring the safety and health of all emergency workers (including firefighters, police, emergency medical service workers, etc.) during large-scale disasters and terrorist attacks is important, yet can be very challenging given unknown risks and competing demands.

Working Youth

- In 2005, there were 54 deaths of workers younger than 18 years of age (BLS, 2007a; NIOSH, 2007i, p. 171).
- In 2003, there were approximately 54,800 visits by youth to emergency departments for work-related injuries. Rates of nonfatal injuries treated in emergency departments generally decrease with age, with rates for youth 15 to 17 years of age exceeded only by rates for workers 18 to 19 years of age (NIOSH, 2007i, p. 171).
- The agricultural industry typically accounts for about 40 percent of young worker fatalities. Most of these deaths occur on family farms. Fatalities among youth working in agriculture are more than 3.5 times higher than for youth working in other industries (Castillo, 2007).
- Child labor laws do not currently apply to children working on family farms or to youth 15 years and older working in agriculture (Castillo, 2007).

a. An assessment of the relevance of the program's activities to the improvement of occupational safety and health.
b. An evaluation of the impact that the program's research has had in reducing work-related hazardous exposures, illnesses, and injuries.

The evaluation committee will rate the performance of the program for its relevance and impact using an integer score from 1 to 5. Impact may be assessed directly (for example, on the basis of reductions in illnesses or injuries) or, as necessary, by using intermediate outcomes to estimate impact. Qualitative narrative evaluations should be included to explain the numerical ratings.

2. Assessment of the program's effectiveness in targeting new research areas and identifying emerging issues in occupational safety and health most relevant to future improvements in workplace protection. The committee will provide a qualitative narrative assessment of the program's efforts and suggestions about emerging issues that the program should be prepared to address.

Using an evaluation framework developed by the National Academies Committee to Review the NIOSH Research Programs (see Appendix A), the committee was to evaluate the relevance, impact, and future directions of the TI Research Program. In addition to an evaluation of what the TI Research Program is producing, the committee was to assess whether the program can be credited with changes in workplace practices, exposures, or occupational injuries or whether such changes or the lack of such changes, are the result of external factors.

The committee was comprised of 10 members and one Framework Committee liaison.[8] The committee had expertise in the areas of occupational health, public health education, medicine, injury prevention and control, epidemiology and biostatistics, labor, industry, program evaluation, ergonomics, and bioengineering. The committee met three times between March and September 2007. Committee subgroup deliberations were held by conference calls prior to the committee's final meeting.

EVALUATION PROCESS

The committee evaluated the TI Research Program for the period from 1996 to 2005, the first decade of NORA. The information evaluated consisted primarily

[8]The Framework Committee liaison is a member of the National Academies Committee to Review the NIOSH Research Programs, which is the committee that developed the framework (see Appendix A of this report) for review of NIOSH research programs.

of materials—organized by traumatic injury goals and subgoals—included in an evidence package provided by NIOSH to the committee. At its first meeting, the committee also heard presentations from NIOSH staff on the overall TI Research Program as well as each of the eight TI Research Program goal areas. A question-and-answer session followed each presentation. IOM project staff also sought the input of stakeholders who had been identified by NIOSH as having an interest in traumatic injury research and prevention. (Additional details on methods and information gathering can be found in Appendix B of this report.)

The committee performed its assessment of program relevance and impact using an evaluation flowchart developed by the Framework Committee (Figure 1-3). The flowchart is based on the NIOSH operational plan of the path from inputs to outcomes for all NIOSH research programs (see Appendix A, Figure 1, p. 148). For each of the eight program goals, the committee conducted a qualitative assessment for each box in the evaluation flowchart. Assessment of relevance was based on the committee's review of program challenges, strategic goals and objectives, inputs, and activities (Boxes A-C), while assessment of impact was based on review of outputs, intermediate outcomes, and end outcomes (Boxes D-F). The Framework Document provides guidance on evidence to be considered in each of these categories. The committee also considered the possible roles of external factors at all steps in the evaluation process. The committee's findings and ratings of overall program relevance and impact are discussed in Chapter 2.

For its assessment of the NIOSH process for targeting new research areas and identifying emerging issues in occupational safety and health, the committee relied primarily on a review of the TI Research Program's strategic goals for the future (a list of these goals was included in the evidence package provided by NIOSH to the committee). A review of these findings, including strengths and weaknesses of the current list of future goals as well as goals to be added or eliminated, is discussed in Chapter 3. The report concludes with a final chapter in which the committee makes recommendations for program improvement. These recommendations are

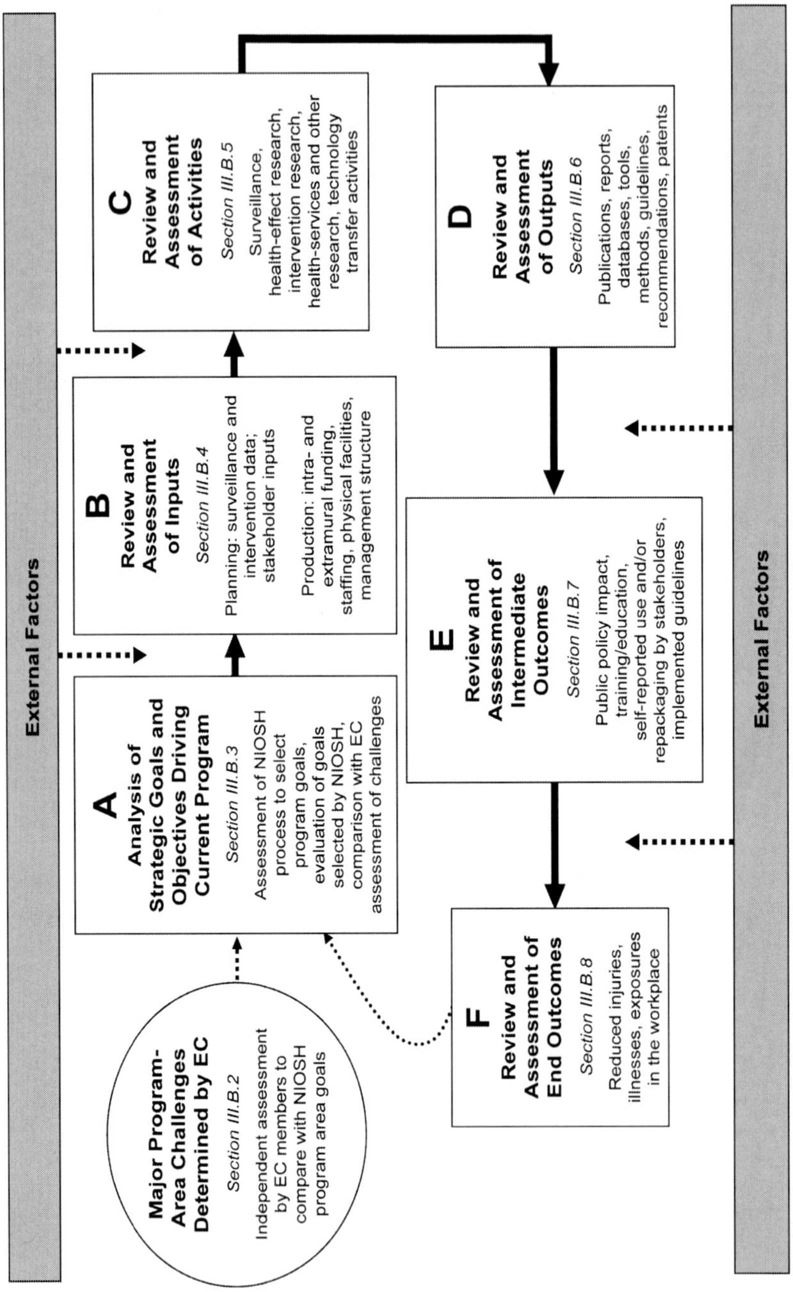

FIGURE 1-3 NIOSH Research Program evaluation flowchart.

not specific to the research goals, but derive from the committee's observations across the goal-oriented reviews.

REFERENCES

Amandus, H. 2007. *Reducing occupational injuries and fatalities due to workplace violence.* PowerPoint presented at the Institute of Medicine's first meeting of the Committee to Review the NIOSH TI Research Program.

BLS (U.S. Bureau of Labor Statistics). 2007a. Census of fatal occupational injuries: Current and revised data. http://www.bls.gov/iif/oshcfoi1.htm#2006 (accessed October 24, 2007).

BLS. 2007b. *News release: Workplace injuries and illnesses in 2006.* Washington, DC: DOL.

Boden, L. I., and A. Ozonoff. 2008. Capture-recapture estimates of nonfatal workplace injuries and illnesses. *Annals of Epidemiology* 18(6):500-506.

Castillo, D. N. 2007. *Reduce injuries and fatalities to working youth.* PowerPoint presented at the Institute of Medicine's first meeting of the Committee to Review the NIOSH TI Research Program.

Dembe, A. E. 2001. The social consequences of occupational injuries and illnesses. *American Journal of Industrial Medicine* 40(4):403-417.

DOL (Department of Labor). 2008. OSHA's mission. http://www.osha.gov/oshinfo/mission.html (accessed July 21, 2008).

Hensler, D. R., M. S. Marquis, A. Abrahamse, S. H. Berry, P. A. Ebener, E. Lewis, E. Lind, R. MacCoun, W. G. Manning, J. Rogowski, and M. E. Vaiana. 1991. *Compensation for accidental injuries in the United States.* Santa Monica, CA: RAND.

HHS (Department of Health and Human Services). 2005. *Budget in brief: FY2005.* http://www.hhs.gov/budget/05budget/fy2005bibfinal.pdf (accessed November 16, 2007).

Hsiao, H. 2007. *Reduce injuries and fatalities due to falls from elevations.* PowerPoint presented at the Institute of Medicine's first meeting of the Committee to Review the NIOSH TI Research Program.

Leigh, J. P., S. Markowitz, M. Fahs, and P. Landrigan. 2000. *Costs of occupational injuries and illnesses.* Ann Arbor, MI: University of Michigan Press.

Morse, T. F., C. Dillon, N. Warren, C. Levenstein, and A. Warren. 1998. The economic and social consequences of work-related musculoskeletal disorders: The Connecticut upper extremity surveillance project (CUSP). *International Journal of Occupational and Environmental Health* 4:209-216.

NIOSH (National Institute for Occupational Safety and Health). 1998. *NORA: Traumatic occupational injury research needs and priorities.* Washington, DC: HHS.

NIOSH. 2007a (unpublished). *Introduction to briefing book.* In the evidence package provided to the Committee to Review the NIOSH TI Research Program. NIOSH.

NIOSH. 2007b (unpublished). *Executive summary of briefing book.* In the evidence package provided to the Committee to Review the NIOSH TI Research Program. NIOSH.

NIOSH. 2007c (unpublished). *Overview of TI Research Program.* In the evidence package provided to the Committee to Review the NIOSH TI Research Program. NIOSH.

NIOSH. 2007d (unpublished). *Current TI research goals and sub goals: Reduce injuries and fatalities due to falls from elevations.* In the evidence package provided to the Committee to Review the NIOSH TI Research Program. NIOSH.

NIOSH. 2007e (unpublished). *Current TI research goals and sub goals: Reduce injuries and fatalities due to machines.* In the evidence package provided to the Committee to Review the NIOSH TI Research Program. NIOSH.

NIOSH. 2007f (unpublished). *Current TI research goals and sub goals: Reduce acute back injury.* In the evidence package provided to the Committee to Review the NIOSH TI Research Program. NIOSH.

NIOSH. 2007g (unpublished). *Current TI research goals and sub goals: Reduce injuries and fatalities among workers in Alaska.* In the evidence package provided to the Committee to Review the NIOSH TI Research Program. NIOSH.

NIOSH. 2007h (unpublished). *Current TI research goals and sub goals: Reduce injuries and fatalities to emergency responders.* In the evidence package provided to the Committee to Review the NIOSH TI Research Program. NIOSH.

NIOSH. 2007i (unpublished). *Current TI research goals and sub goals: Reduce injuries and fatalities to working youth.* In the evidence package provided to the Committee to Review the NIOSH TI Research Program. NIOSH.

Pegula, S. 2004. Fatal occupational injuries at road construction sites. *Monthly Labor Review* December:43-47.

Pransky, G., K. Benjamin, C. Hill-Fotouhi, J. Himmelstein, K. E. Fletcher, J. N. Katz, and W. G. Johnson. 2000. Outcomes in work-related upper extremity and low back injuries: Results of a retrospective study. *American Journal of Industrial Medicine* 37(4):400-409.

Reville, R. T., L. I. Boden, J. E. Biddle, and C. Mardesich. 2001. Comparing New Mexico PPD outcomes with PPD outcomes in other states. In *An evaluation of New Mexico workers' compensation permanent partial disability and return to work.* Arlington, VA: RAND Corporation.

Rosenman, K. D., A. Kalush, M. J. Reilly, J. C. Gardiner, M. Reeves, and Z. Luo. 2006. How much work-related injury and illness is missed by the current national surveillance system? *Journal of Occupational and Environmental Medicine* 48(4):357-365.

Sengupta, I., V. Reno, and J. F. Burton, Jr. 2007. *Workers' compensation: Benefits, coverage, and costs, 2005.* Washington, DC: National Academy of Social Insurance (NASI).

Silverstein, B. A., D. S. Stetson, W. M. Keyserling, and L. J. Fine. 1997. Work-related musculoskeletal disorders: Comparison of data sources for surveillance. *American Journal of Industrial Medicine* 31(5):600-608.

Smith, G. S., H. M. Wellman, G. S. Sorock, M. Warner, T. K. Courtney, G. S. Pransky, and L. A. Fingerhut. 2005. Injuries at work in the U.S. adult population: Contributions to the total injury burden. *American Journal of Public Health* 95(7):1213-1219.

Strunin, L., and L. I. Boden. 2004. Family consequences of chronic back pain. *Social Science and Medicine* 58(7):1385-1393.

2

Assessment of Programmatic Relevance and Impact

This chapter constitutes the committee's response to the first part of the charge to assess the National Institute of Occupational Safety and Health (NIOSH) Traumatic Injury (TI) Research Program's contribution—through occupational safety and health research—to reductions in workplace hazardous exposures, illnesses, or injuries through

- An assessment of the relevance of the program's activities to the improvement of occupational safety and health; and
- An evaluation of the impact that the program's research has had in reducing work-related hazardous exposures, illnesses, and injuries.

This chapter reviews the inputs, activities, outputs, and outcomes of each of the TI Research Program's eight goals. Much of the evidence derives from the evidence package prepared by TI Research Program staff. While the committee had information on extramural TI research grants, the information in the evidence package was presented with a focus on the work of the staff in the Division of Safety Research (DSR). Thus, the committee's evaluation is primarily of the TI Research Program's intramural research and related efforts. The committee considered the TI Research Program under review to include only those activities, outputs, and outcomes included in the evidence package and evaluated the program on that basis. The committee did not attempt to independently verify the facts in the package (for example, a statement of how many brochures were mailed was believed to be accurate), nor did the committee seek to expand the evidence by searching outside the evidence package. Although the committee was aware of some NIOSH-sponsored work that did not appear in the evidence package but that the committee would have thought part of the TI Research Program, the committee decided to evaluate only what was presented to them. The committee did use its expert judgment and knowledge of the field to evaluate claims in the evidence package about the role of NIOSH-funded work in intermediate and end outcomes. The committee did not attempt to read and evaluate the quality of each research project or each dissemination product. The committee's evaluation was informed by the evidence package, committee expertise in the field, stakeholder input, and published material. This chapter begins with a section describing the external factors that impact the work the TI Research Program undertakes, as well as the outcomes one might reasonably expect. This is followed by a section describing issues, methods, and challenges in occupational injury surveillance, which cuts across all eight program goals.

EXTERNAL FACTORS WITH BROAD IMPACT ON THE TI RESEARCH PROGRAM

External factors "may be considered as forces beyond the control of NIOSH that may affect the evolution of the program" (see Appendix A, p. 137). They may influence NIOSH research at any phase. External factors identified by the committee that broadly impact NIOSH research activities, including those of the TI Research Program, include budgetary issues (such as congressional funding and earmarks), lack of complete occupational injury surveillance data, inadequate action by regulatory agencies such as the Occupational Safety and Health Administration (OSHA), the sharing of research responsibility with other agencies, and industry challenges; each of these is discussed below. External factors also impact the TI Research Program's work in specific goal areas. Some of these are discussed in the sections that appear later in this chapter outlining each of the eight goals.

Budgetary Issues

NIOSH was appropriated \$286 million in fiscal year (FY) 2005 to carry out its mission as the sole federal agency responsible for conducting research for the prevention of occupational injury and illness (NIOSH, 2007a, p. 17). The total TI Research Program budget for all eight research goals in FY2005 was \$17.2 million (see Table 1-2), an amount the committee finds to be inadequate given the scope of the TI Research Program's mandate. More recently, the NIOSH budget has decreased; it was appropriated \$255 million and \$253 million for research in FY2006 and 2007, respectively (NIOSH, 2007a, p. 17). The FY2006 decrease reflected a new

Centers for Disease Control and Prevention (CDC) budget structure that draws from program allocations to support operational (overhead and indirect) costs. According to the evidence package, in 2006 Congress redirected $35 million from the NIOSH budget appropriation to CDC for business support services (NIOSH, 2007a, p. 17).

Congressional funds received by the TI Research Program are sometimes earmarked for a specific purpose, such as research that targets a specific industry or population. Such mandates have led to TI Research Program initiatives and ensured sustained research funding in several specific areas such as children in agriculture, workers in Alaska, firefighters, and workplace violence (NIOSH, 2007b, p. 47). Congress or the Department of Health and Human Services (HHS) can also assign NIOSH the task of leading or participating in research on evolving public health issues (NIOSH, 2007a, p. 17). For example, after the September 11, 2001, terrorist attacks, Congress and HHS directed NIOSH to evaluate the health of workers and volunteers at Ground Zero. These targeted areas can be productive, but the strategic decision of where to focus funds and effort is sometimes external to NIOSH and its own strategic thinking and analysis.

Incomplete Occupational Injury Surveillance Data

As will be discussed in greater detail in the following section, surveillance data on occupational injuries is incomplete, particularly for nonfatal injuries, due to factors such as gaps in the scope of surveillance and poor reporting of injuries in national surveys. Far more nonfatal occupational injuries than fatal occupational injuries occur each year (BLS, 2008), and risk factors for fatal and nonfatal injuries are not necessarily the same. Fatal injuries are not surrogates for nonfatal injuries. Yet the TI Research Program has had to rely primarily on fatality data to inform its research efforts.

Key variables missing from data sets also interfere with the TI Research Program's ability to do research in particular goal areas. Examples cited in the evidence package include lack of work relationship information in current crash data systems, which impedes the identification of risk factors for occupational crashes (NIOSH, 2007c, p. 58) and the lack of accurate data on the number of active firefighters (career and volunteer) and the number of hours of exposure, which precludes calculation of accurate firefighter fatality rates (NIOSH, 2007d, p. 154).

Inadequate Regulatory Action

The impact of TI Research Program activities is limited by NIOSH's mandated inability to regulate and enforce measures that impact the occurrence of occupational injuries. According to the Occupational Safety and Health Act (OSH Act) of 1970, NIOSH is responsible for conducting research and making recommendations for occupational safety and health standards. Criteria Documents are developed to provide the basis for comprehensive occupational safety and health standards. These documents generally contain a critical review of the scientific and technical information available on the prevalence of hazards, the existence of safety and health risks, and the adequacy of methods to identify and control hazards. Special Hazard Reviews (SHRs) and Occupational Hazard Assessments are written to complement NIOSH recommendations for standards. These documents assess safety and health problems and recommend appropriate methods for control and monitoring.[1]

Although OSHA's work is informed by NIOSH research and NIOSH research can affect change in the workplace absent OSHA regulation, OSHA—in the Department of Labor (DOL)—is the agency ultimately responsible for the development and enforcement of workplace safety and health regulations.[2] Regulatory inaction on the part of OSHA can delay the translation of TI Research Program findings into workplace policies. Additionally, depending on the research area, inaction or underachieved expectations on the part of regulatory agencies other than OSHA (e.g., the Department of Transportation or the U.S. Department of Agriculture [USDA] or other agencies) can also stall translation of TI Research Program findings. In the absence of regulations, employers may be less likely to adopt safety measures.

Exemptions from federal occupational safety and health regulations—such as the exemption from the 1970 OSH Act of employers of public-sector workers, the congressional exemption for farms employing fewer than 11 people, and the family farm exemption in federal child labor laws—are also barriers to application of some of NIOSH's research findings.

Sharing Research Responsibility with Other Agencies

Garnering support for research where there is a shared responsibility can be complicated when the occupational component represents a relatively small proportion of the overall societal problem. For example, this has been true for NIOSH in the area of injuries due to motor vehicle incidents; although NIOSH focuses solely on occupation-related crashes, it still shares research responsibility with the

[1]During the time period of the committee's evaluation, NIOSH has issued four Criteria Documents and one SHR. The last Criteria Document was issued in 2006; the previous was issued in 1998. Only the 1997 SHR, Child Labor Research Needs, addressed a TI concern.

[2]OSHA and the Mine Safety and Health Administration have delegated approval of respirators to NIOSH.

National Highway Traffic Safety Administration (NHTSA). Occupation-related crashes represent only about 3 percent of the total number of motor vehicle crashes (NIOSH, 2007c, p. 58), a fact that has made it difficult for the TI Research Program to synergistically and strategically engage the broader traffic safety community and also results in the exclusion of some occupational risk factors from national policy (e.g., exclusion of ambulances from crashworthiness consideration). To some extent the same has also been true for research on workplace violence, because violence that occurs in the workplace is a comparatively small part of the overall problem of violence.

Sharing research responsibility with other agencies can also make causal attribution of end outcomes to the TI Research Program's research findings more complicated (e.g., decreases in occupationally related motor vehicle deaths and workplace homicides can be reflections of an overall change in motor vehicle deaths and homicides rather than occupational initiatives).

Industry Challenges

Some challenges to conducting research are related to the nature of the industries targeted by the TI Research Program—for example, the mobile nature of the agriculture and construction industries and the small-business composition of the retail industry. It is difficult to enlist employer participation in studies to prevent rare events. There are also barriers to corporate investment in injury prevention technologies that are related to there being no accepted way of predicting how improvements in workplace safety will reduce the risk of worker injury.

ORGANIZATIONAL DESIGN

Strategically and organizationally, the TI Research Program supports a larger NIOSH mission. In terms of organizational design, the TI Research Program has been an instrumental part of the crosscutting NIOSH Program Portfolio Matrix Management Initiative (Table 1-1). NIOSH's formal organizational structure depicts DSR as part of a product/service organizational structure with geographic distribution. Its Matrix Management Initiative has recently been reorganized to be industrial-sector programs × crosscutting programs,[3] of which the TI Research Program is a component.

The traditional benefits of matrix management include the flexibility derived from resourcing from both axes of the matrix. The TI Research Program goals

[3]As described in Chapter 1, the coordinated emphasis areas have been moved into the crosscutting programs.

represent a mix of risk factors in different occupations, various work environments, and different vulnerable populations. Since goals are in crosscutting categories and NIOSH is organized in a matrix, the assumption is that the structure (matrix) will support the strategy of crosscutting research, but previous users of matrix organizational structures caution that the structure can also be a source of overlap and inefficient use of resources. As an example, "fall from heights," a major source of occupational injury and fatality, is a major problem across sectors (construction falls from rooftops, surface mining falls from heavy equipment, falls from towers in telecommunications, etc.). Falls may be experienced disproportionately by some worker groups (e.g., Latinos) as well as in certain firms (e.g., small). Given the relatively modest budget for occupational health and safety research in general, the organizational structure should support information sharing and efficient use of resources in such cases, rather than having multiple, parallel efforts to address a common issue. Of particular concern is whether the crosscutting programs and the sector programs are maximizing the efficient use of resources and information.

Matrix management is an attempt to provide flexibility in terms of personnel assignment and reassignment and in a sense, an attempt to allow the benefits of the two axes while promoting synergistic additional benefits. In terms of injury prevention, there is great potential. To illustrate the potential of crosscutting and true matrices, consider Haddon's matrix (Haddon, 1968, 1972). Combining his medical and engineering expertise, Haddon provided a "matrix" for analyzing injury based on the person injured, the cause of the injury, and the environment and for identifying temporal factors across the pre-event, event, and post-event life cycle. To further investigate the root cause of falls from heights discussed earlier, kinetic energy is closer to the root cause and, in fact, is the most common agent that causes occupational injury (e.g., tissue tearing, crushing with bleeding, organ failure and death from car crashes, falls, gunshot wounds). While prevention, modification, or attenuation of the release of this energy is common across all of the sectors that NIOSH has created, the TI Research Program (as well as many or most others) may be stuck between sectors, simply because of the way NIOSH is organized. One potential solution—if reorganization is not desired or practical—is to use something like Haddon's matrix as what is called in the organizational field an "integrating mechanism" that facilitates the organization and integration of common information.

In summary, there are common pitfalls with matrix management that should be noted as the TI Research Program and NIOSH continue to evolve. Internally, there can be confusion over lines of authority for personnel attached to multiple programs or units within the organization. Although this classic challenge was not observed in the committee's review, the related issue of project "ownership" and associated information sharing was observed. That is, it was sometimes difficult to fully understand which projects fell under the purview of the TI Research Program, because of its positioning within a matrix structure. The committee is an external body to which NIOSH needed to communicate its framework structure and processes, and committee members were occasionally confused about the TI Research Program's responsibilities within that framework. As NIOSH continues to work externally, for example, in sharing responsibility with other agencies, it will have to communicate effectively how it is structured organizationally (i.e., product/service), how it is managed (matrix), and how it operates (logic model), and more importantly, it should effectively share and manage its portfolio of injury prevention knowledge in a resource efficient manner (e.g., effective integrating mechanisms).

SURVEILLANCE

Surveillance of traumatic injuries provides the empirical basis for setting research priorities and for evaluating the impact of research, knowledge transfer, and intervention activities. Although substantial strides have been made over the past 20 years in the development of a national surveillance system for fatal occupational injuries, there is widespread evidence of substantial disparities between the number of nonfatal occupational injuries that are reported and the actual number that occur.

A series of studies beginning in the 1980s demonstrated that both nonfatal and fatal injuries are underreported (Azaroff et al., 2002; Stout and Bell, 1991). Based on these studies—many of which were performed or supported by NIOSH—the TI Research Program developed the NTOF (National Traumatic Occupational Fatality) Surveillance System. This system provided much improved reporting of injuries resulting in death and contributed to the creation of the Census of Fatal Occupational Injuries (CFOI) in 1992. Improved surveillance of fatal occupational injuries is an impressive achievement of the TI Research Program and is currently an important input into the TI Research Program's planning process.

The TI Research Program also expanded fatality surveillance by establishing the Fatal Assessment Control and Evaluation (FACE) Program.[4] This program, conducted in collaboration with state partners, provides in-depth information about

[4]The NIOSH FACE Program concentrates on investigations of fatal occupational injuries. It has two components—an in-house program that began in 1982 in which participating states voluntarily notify NIOSH of traumatic occupational fatalities resulting from targeted causes of death (currently, deaths associated with machinery, deaths of youths under 18 years of age, and street/highway construction work zone fatalities) and a state-based program that began in 1989 in which nine states with cooperative agreements with NIOSH conduct surveillance, targeted investigations, and prevention activities using the FACE model (NIOSH, 2008). The number of state-based FACE programs has declined in recent years due to lack of funding despite a growing interest and expertise on the part of the states.

targeted types of deaths. This information—collected through onsite, research-oriented field investigations of fatal incidents—is used to develop recommendations to prevent future deaths. These recommendations are in turn disseminated widely to industry and labor professionals, manufacturers, policy makers, and health and safety professionals.

Although substantial strides have been made in the surveillance of fatal occupational injuries, many types of traumatic workplace injury rarely result in death, and fatal injuries are not good surrogates for nonfatal injuries. Planning and evaluation also depend on accurate surveillance of nonfatal injuries. It is widely recognized that the current major source of national surveillance data on nonfatal occupational injuries—the Bureau of Labor Statistics (BLS) Annual Survey of Injuries and Illnesses—is very incomplete. National estimates of injuries derived from this survey exclude approximately 22 percent of the U.S. workforce, including workers in the public sector and the self-employed (Leigh et al., 2004). There is also mounting evidence of injury underreporting on the OSHA-required injury logs maintained by employers on which the survey is based (Conway and Svenson, 1998; Rosenman et al., 2006). Moreover, completeness of reporting likely varies by type of injury, employer characteristics, and worker characteristics, although due to the nature of the problem the extent of variability is unknown.

Gaps in surveillance and incomplete reporting result in substantial underestimates of the national burden of occupational injuries. As a consequence, decisions based on these data may allocate fewer resources to research and interventions designed to reduce workplace hazards. Secular changes in reporting over time can (if reporting improves) suggest that successful programs have failed or (if reporting declines) suggest that ineffective programs have succeeded. In addition, differential reporting by condition, subpopulation, or employer type can lead to focusing prevention efforts in the wrong areas. Injuries for which we have particularly poor measures of incidence may receive inadequate public health attention.

The TI Research Program has carried out substantial surveillance activities and research in order to provide more reliable surveillance data, including data on nonfatal injuries. Perhaps the most innovative of these efforts has focused on agriculture and young workers. The Child Agricultural Injury Program, for example, has developed several new surveillance activities designed to provide an ongoing picture of the incidence of injuries among both adults and youths. For adult farm workers, Occupational Injury Surveillance of Production and Agriculture (OISPA) surveys farm operators to estimate the number and incidence of occupational injuries to adults working on farms. In addition, NIOSH has developed a farm worker injury module to supplement data already collected from workers by the DOL in the National Agricultural Workers Survey (NAWS). This survey includes both adult and young workers. Using ongoing survey relationships of the National Agricultural

Statistics Service (NASS), NIOSH has collected information from farm operators about agricultural injuries to youth in 1998, 2001, and 2004 (the Childhood Agricultural Injury Surveys [CAIS]). A parallel survey of minority farms, the Minority Farm Operator Childhood Agricultural Injury Surveys (M-CAIS), was conducted in 2000 and 2003. These surveys include a project that ascertained the number of agricultural injuries incurred by ninth to twelfth graders in Minnesota as reported by emergency departments.

Since the early 1990s, NIOSH has had an interagency agreement with the Consumer Product Safety Commission to collect data on nonfatal occupational injuries through the National Surveillance of Nonfatal Occupational Injuries Using the National Electronic Injury Surveillance System (NEISS). NEISS collects data on injuries resulting in emergency department visits from a sample of U.S. hospitals. Not all injuries are seen in emergency departments, but the agreement has allowed NIOSH to collect data on nonfatal injuries for all industries and occupations and has allowed for better understanding of risk factors, particularly for those injuries that are more likely to require emergency department visits. Through the agreement, NIOSH has also been able to conduct follow-up surveys of injured workers to collect more detailed information on their injuries.

To describe risks to working youth, NIOSH has used NEISS and developed new surveillance sources both by building on existing surveys conducted by other agencies and by funding new surveillance activities. These include new state-based surveillance activities in Wisconsin and Massachusetts focused on working youth.

Despite these activities, there remains a paucity of high-quality national data on nonfatal injuries. This may be one reason that the TI Research Program frequently seems to focus on fatality data in setting program priorities. Although fatal injuries should carry substantial weight, too much emphasis on fatalities can lead to an underinvestment in research and transfer activities relevant to injury types that are very rarely fatal yet impose significant human and economic costs.

Notably, occupational injury surveillance is dependent upon having valid information about the number of injuries that occur. It also requires information about populations at risk (denominator data) necessary to generate injury rates that allow for identification of disproportionate risks among segments of the population and subsequent priority setting. Better information on employment and hours at work (as well as injuries) is also needed, especially for the vulnerable populations such as youth and immigrant and minority workers whose work experience may not be well captured in traditional sources of employment data. Optimally, information on worker exposure to hazards would be available to allow for exposure-based estimates of risk. NIOSH, in concert with other agencies, can work to develop better systems for capture of denominator data.

In summary, NIOSH has engaged in a range of activities, both intramural and extramural, that have made strong contributions to improving surveillance. Still, these projects do not appear to be part of a coordinated interagency strategy to improve national surveillance of traumatic nonfatal occupational injuries. For example, there is little research funded to identify the strengths and weaknesses of current national systems or to find ways of combining data to generate better estimates of injury incidence. NIOSH has used population-based surveillance data, but more work could be performed to identify the strengths and weaknesses of new survey-based systems or to attempt to compare or combine data. In its future TI Research Program activities, NIOSH would benefit from an overall strategy of surveillance research and implementation designed to lead to a better national nonfatal injury surveillance program. This might be considered as a separate goal.

GOAL 1: REDUCE INJURIES AND FATALITIES DUE TO MOTOR VEHICLES

NIOSH and the TI Research Program's involvement in motor vehicle injury research began in the late 1990s in response to data demonstrating a lack of progress in reducing work-related crashes and fatalities. At this time, TI Research Program staff also recognized the problem of poor data capture tying crash incidents to occupational injuries in crash data systems. No one had yet taken the lead in making improvements in this area (NIOSH, 2007c, p. 53).

The committee supports NIOSH's decision to become a key player in motor vehicle injury research. There is a niche for NIOSH to evaluate aspects of motor vehicle injuries and workplace intervention strategies that are not already researched by other agencies.

NIOSH developed two subgoals for research aimed at reducing injuries and fatalities due to motor vehicles:

1.1. Reduce occupational injuries and fatalities due to highway motor vehicle crashes[5]
1.2. Reduce occupational injuries and fatalities due to motor vehicle incidents in highway and street construction work zones

Planning and Production Inputs

Planning inputs were comprehensive and supported the need for research in both subgoal areas. Inputs were both qualitative and quantitative.

[5]Subgoal 1.1 specifically targeted (1) professional truck drivers and (2) workers who drive or ride in motor vehicles during work-related travel but who are not professional drivers (NIOSH, 2007c, p. 54).

Subgoal 1.1 inputs included fatal and nonfatal crash injury data from the BLS, which allowed identification of important crash risk factors for truck drivers. A qualitative source of input was feedback from stakeholders at conferences and meetings, which revealed agreement on a need for additional surveillance and risk factor research for truck drivers. The TI Research Program's research on workers who drive or ride in motor vehicles during work-related travel, but who are not professional drivers, was initiated to address a research and regulatory gap (NIOSH, 2007c, pp. 54-55).

An important input into the TI Research Program's work on subgoal 1.2 was a review of BLS data, which showed that most fatalities occurring at road construction sites between 1995 and 2002 were reported to be transportation-related incidents (NIOSH, 2007c, p. 62). There had also been concern among stakeholders that increases in worker injuries would occur in coming years due to increases in road construction and structural changes in the industry, making this a potentially emerging issue (NIOSH, 2007c, p. 62).

Combined intramural–extramural funding for motor vehicle research for the period FY1997-2005 totaled $12,657,927. Only the back injury program had lower total funding for this period. With the exception of a small funding decline in FY2003, intramural funding for motor vehicle research increased each year from FY1997 to FY2005, reflecting the growth of this program over time. However, the proportion going to extramural research began to decline in FY2003, and in FY2005, it accounted for less than 1 percent of the total funding for motor vehicle research (see Table 1-2). Full-time equivalents (FTEs) increased from a low of 4.3 in FY1998 to 14.8 in FY2005.

Activities

Activities for subgoal 1.1 included surveillance research as well as research to identify crash risk factors and interventions. Perhaps owing to the fact that NIOSH and the TI Research Program are relatively new to motor vehicle safety research, activities for subgoal 1.1 were rather limited given the broad scope of the goal. The committee agrees with the TI Research Program that it is continuing to develop a niche within transportation safety research. Appropriately, the TI Research Program has been working with partners such as the American Society of Safety Engineers and the Network of Employers for Traffic Safety to learn how it can best serve employers and employee drivers through research and outreach (NIOSH, 2007c, p. 55). Having focused motor vehicle safety research goals (rather than the broader goal of reducing injuries and fatalities due to highway motor vehicle crashes) will be important for informing the TI Research Program's activities going into the future.

For subgoal 1.2, which the committee finds to be an excellent niche for the TI Research Program, highway and construction work zones were added as a target for FACE investigations by the program in 1999. TI Research Program staff has since completed 12 FACE investigations involving transportation-related incidents in highway and street construction zones (NIOSH, 2007c, p. 63), and the state-based FACE program has conducted investigations of more than 50 highway work zone deaths (NIOSH, 2008). FACE investigations have been for worker fatalities from motor vehicles used for construction, as well as worker fatalities from passing motor vehicle traffic. These investigations represent a significant contribution to the knowledge of factors potentially contributing to work zone deaths.

From 2000 to 2002, the TI Research Program's work for subgoal 1.2 was focused largely on intervention research that tested the effectiveness and reliability of proximity warning systems (PWS) in alerting equipment operators and workers when workers enter blind areas around road construction site equipment. The focus on this intervention is appropriate given the data demonstrating that most fatalities occurring at road construction sites between 1995 and 2002 involved workers being struck by a vehicle or mobile equipment (NIOSH, 2007c, p. 62). In 2002, the TI Research Program began testing some PWS devices, as well as an internal traffic control plan (ITCP)[6] on active construction road sites. Neither the PWS devices nor the ITCP have yet been proven effective (NIOSH, 2007c, p. 63), but additional research is planned to evaluate their effectiveness on highway construction sites (NIOSH, 2007c, p. 65). The committee supports the TI Research Program's decision to continue research in this area for eventual work site implementation.

Some diffusion and dissemination research has been performed for subgoal 1.2 in the form of evaluations of effectiveness of safety training programs for Spanish-speaking road construction workers. This is an important area for research because of the large and growing number of Spanish-speaking construction workers. The NIOSH Education and Information Division is heading this effort (NIOSH, 2007c, p. 63).

Transfer Activities

The TI Research Program has engaged in proactive transfer activities for both motor vehicles subgoals.

A planned transfer effort is demonstrated for subgoal 1.1, where TI Research Program staff developed dissemination strategies for NIOSH publications on motor vehicle safety. More than 120,000 of these publications have been distributed to targeted distribution lists and at conferences. The TI Research Program also

[6] ITCPs represent an administrative approach designed to assist in controlling construction vehicle and worker movements inside the work zone (NIOSH, 2007c, p. 63).

prepared and disseminated a Hazard Review discussing identification of worker groups at high risk of crashes, among other injuries. Press releases were used to increase the visibility of some NIOSH and NIOSH-funded projects related to motor vehicle safety (NIOSH, 2007c, pp. 56-57).

For subgoal 1.2, approximately 21,000 copies of the NIOSH publication "Building Safer Highway Work Zones: Measures to Prevent Worker Injuries from Vehicles and Equipment"[7] were distributed through targeted mailings and handouts at conferences and exhibitions. This document was authored by TI Research Program staff and outlines measures for reducing injuries in highway work zones (NIOSH, 2007c, p. 63).

Other transfer activities involving dissemination have been more passive. The NIOSH publication "Building Safer Highway Work Zones: Measures to Prevent Worker Injuries from Vehicles and Equipment"[8] and the FACE investigation reports are available for download from the NIOSH Web site. Technical assistance in the form of diagrams of blind areas for work zone equipment is distributed externally upon request (NIOSH, 2007c, pp. 63-64).

TI Research Program staff involvement on the American National Standards Institute (ANSI) Z15 committee is an important collaborative transfer activity. The ANSI Z15 committee developed ANSI standard Z15.1-2006, which outlines minimum standards for workplace traffic safety training programs. The standard impacts employers whose employees drive for work. TI Research Program staff contributed to all parts of the standard and led the development of the portion of the standard on crash data collection and incident analysis (NIOSH, 2007c, p. 55).

Outputs

The TI Research Program for motor vehicles has produced a number of outputs, including FACE investigations, peer-reviewed journal articles, and NIOSH publications, and has sponsored workshops and conferences. The TI Research Program does not appear to have generated new knowledge per se related to subgoal 1.1. Instead, outputs have focused on describing risks and new applications of previously used research methods. There do not appear to be widely cited peer-reviewed publications of breakthrough results. Outputs for subgoal 1.2—in particular the TI Research Program-authored NIOSH report "Building Safer Highway

[7]Pratt, S. G., D. E. Fosbroke, and S. M. Marsh. 2001. Building safer highway work zones: Measures to prevent worker injuries from vehicles and equipment. HHS (NIOSH) Publication No. 2001-128.

[8]See footnote 7 for reference.

Work Zones: Measures to Prevent Worker Injuries from Vehicles and Equipment"[9] and the 13 highway work zone FACE investigations—are more innovative.

Outcomes

TI Research Program staff's contribution to the development of the ANSI Z15.1 standard is an important intermediate outcome. The standard will likely lead to workplace change, especially for small businesses, through the guidance it provides on how to protect the safety of workers who drive as part of their work. Approval of this standard is a noteworthy accomplishment because the Federal Motor Carrier Safety Regulations apply only to commercial drivers, leaving many small- and medium-sized carriers without guidance for establishing or carrying out a motor vehicle safety program. Although TI Research Program staff participation in standards setting and other consensus activities is necessary and is recognized as such by the committee, it is difficult to know what proportion of the success of the ANSI standard was the result of the TI Research Program staff's participation in its development.

Another very important intermediate outcome was the incorporation of a NIOSH recommendation related to Hazardous Occupations Order (HO) No. 2 (motor vehicle occupations) into a final rule published by the DOL in 2004. NIOSH's recommendation was that the HO be changed to comply with 1998 congressional amendments to the Fair Labor Standards Act (FLSA) (NIOSH, 2007c, p. 57), which outlined new rules for teenagers who drive for work (the Teen Drive for Employment Act [P.L. 105-334]). The DOL adopted the recommendation and published a final rule[10] on the HO, which prohibits all workplace driving on public roadways by 16-year-olds and places restrictions on workplace driving by 17-year-olds.

There are other indicators of stakeholder use of the TI Research Program's work on work zone safety. The NIOSH publication "Building Safer Highways and Work Zones: Measures to Prevent Worker Injuries from Vehicles and Equipment,"[11] for example, has been used by stakeholders for a variety of purposes and has resulted in workplace changes to safety training. Key measures from this document have been incorporated into course materials for an OSHA training course provided to member construction companies for the National Safety Council and the Ameri-

[9]See footnote 7 for reference

[10]Regulations implementing the act permit the Secretary of Labor to prohibit the employment of youth in occupations declared "particularly hazardous for the employment of children . . . or detrimental to their health or well-being" (29 USC 201 Sec. 3(1)). These prohibited activities are referred to as Hazardous Occupations Orders (HOs). The minimum age, by statute, for HOs in nonagricultural occupations is 18; the minimum age in agricultural occupations is 16.

[11]See footnote 7 for reference.

can Road and Transportation Builders Association. This document has been used by a variety of other organizations (universities, an insurance company, and the Federal Emergency Management Agency, among others) as well for purposes such as safety training and strategic planning, informing best-practice guidelines, and the development of risk management recommendations (NIOSH, 2007c, p. 64). Stakeholders have also expressed interest in the diagrams of blind areas around work zone equipment. These reports have been requested and used by one state's health department for the development of recommendations for internal traffic control, by PWS device manufacturers for product development and marketing, and by construction companies for safety training (NIOSH, 2007c, p. 65).

Discussion

The TI Research Program is to be commended for recognizing a decade ago that motor vehicles are a major contributor to occupational injury and death and it has an important role in motor vehicle injury research.

As stated previously, NIOSH and the TI Research Program are continuing to find their niche within transportation safety research and are collaborating with partners to identify how research can best meet employer and employee work-related motor vehicle safety concerns. It is perhaps for this reason that activities for subgoal 1.1 were rather limited for the evaluation period. The committee found research activities for subgoal 1.2 on work zone safety to be highly relevant and a good use of the TI Research Program's limited resources to make a contribution to a focused area not researched by other agencies. A discussion of possible future motor vehicle injury research areas for the TI Research Program is included in Chapter 3 of this report.

Both subgoals showed a broad array of transfer activities. The technical assistance provided by TI Research Program staff to the ANSI standards committee is particularly admirable. Although activities for subgoal 1.1 were not particularly innovative, many important outputs resulted from this work. The approval of the ANSI Z15.1 standard and the incorporation of the NIOSH recommendation to revise the FLSA section on child labor are strong intermediate outcomes impacting safety training and regulations for workplaces.

GOAL 2: REDUCE INJURIES AND FATALITIES DUE TO FALLS FROM ELEVATIONS

NIOSH developed three subgoals for its research aimed at reducing falls from elevations:

2.1. Reduce worker falls from roofs
2.2. Improve fall-arrest harnesses
2.3. Reduce worker falls from telecommunications towers

Planning and Production Inputs

Quantitative inputs to the TI Research Program's work on falls from elevations were BLS data on the number of fatal and nonfatal occupational injuries resulting from falls from elevations as well as information on societal costs. The construction industry was identified as having the highest frequency of falls from elevations—most often from roofs, ladders, and scaffolds. OSHA regulations require the use of harnesses, guardrails, or safety nets for tasks that are performed above 6-feet height (NIOSH, 2007e, p. 69). A comprehensive literature review identified factors and information gaps related to subgoal 2.1. A lack of current anthropometric data relevant to harness design was noted in the establishment of subgoal 2.2 (NIOSH, 2007e, p. 75). Subgoal 2.3 is shaped by statistics showing that the fatality rate from falls in the telecommunications industry is as much as a hundredfold higher than the average fatality rate across all industries. A doubling of fatalities between 2005 and 2006 is further justification for attention to this area (NIOSH, 2007e, p. 78).

The total intramural and extramural combined funding for this project between FY 1997 and FY 2005 was $14,449,742 ($12,148,782 intramural and $2,300,960 extramural). The funding amount seems to be fairly consistent over all 9 years, ranging from $1,136,996 in FY 1999 to $1,853,368 in FY 2003. One noteworthy change is that in FY 2005, the extramural funding was reduced to only $250, which may indicate that NIOSH will phase out extramural funding in this area altogether. FTEs ranged from a low of 8.0 in FY 1998 to a high of 15.4 in FY 2000 (Table 1-2).

Overall, the committee found the subgoals to be appropriate for the overarching challenge of reducing falls from elevations. The committee was surprised, however, to see no formal recognition by the TI Research Program of slips, trips, and falls from the same elevation. Although work-related slips, trips, and falls from the same elevation do not result in as many fatalities as falls from heights, they do occur more often and are a significant source of morbidity and costs.

Activities

The TI Research Program has taken a strong engineering approach in this area. For subgoal 2.1, it developed new technologies and used its virtual reality lab to better understand balance and sensory cues. Specifically, the TI Research Program developed an adjustable guardrail, validated virtual reality technology for fall prevention research, identified the effects of visual cues on balance control, developed improved footwear designs, established sensory-enhancing technology to improve worker balance, and developed safer scissor lifts and work practices. The TI Research Program also used virtual reality technologies and sensory enhancing technologies and developed safer scissor lifts (NIOSH, 2007e, pp. 70-71).

To improve fall-arrest harnesses, the TI Research Program used advanced scanning technology in a pilot study to perform full-body scans of actual workers in actual work positions (standing and suspended). It then demonstrated that the current criteria for fit were inappropriate for two harnesses. In so doing, the TI Research Program demonstrated the need for updated harnesses. Working with harness manufacturers, it updated mathematical parameters that were applied to a much larger anthropometry database in order to enable determinations of necessary adjustments to harness components. The TI Research Program also determined that another of its inventions—a harness accessory that would right a fallen worker to a sitting position—lengthened the time a suspended fallen worker could endure without physiologic stress (NIOSH, 2007e, pp. 75-76).

Activities for subgoal 2.3 were FACE investigations of 12 telecommunications tower-related fall fatalities that occurred from 1992 to 2001. The leading causal factors for these falls were determined to be failure to use or improper use of personal protective equipment; improper, inadequate, and improper maintenance of hoisting equipment; and inadequate employer safety programs and training. The TI Research Program also reviewed BLS and OSHA data to identify and characterize fatal falls that occurred from 1992 to 2005 among workers constructing or maintaining telecommunication towers (NIOSH, 2007e, p. 78). This work led to transfer activities and outputs, which are described below.

Transfer Activities

The TI Research Program has engaged in significant transfer activities for its research on falls. Specifically, the program has been successful at generating the interest of potential manufacturers in several technologies. A patent was filed for the guardrail assembly and the TI Research Program held a workshop to engender interest in the use of the new system (NIOSH, 2007e, p. 70). Two companies have expressed interest in the product and are conducting product development and cost analyses. The virtual reality technology developed by the TI Research Program is of interest to other countries, which would greatly extend this work. The TI Research Program has signed a letter of agreement with a scissor lift manufacturer to jointly develop aerial lift designs. The partner is testing TI Research Program-developed software to determine its applicability to new lift design (NIOSH, 2007e, p. 77).

With regard to subgoal 2.2, the TI Research Program has initiated efforts with two major harness manufacturers that account for more than half of the national market share to incorporate the anthropometric findings into more appropriate harnesses. For subgoal 2.3, the TI Research Program has led the transfer of findings to workers and companies through the dissemination of FACE reports through National Association of Tower Erectors (NATE) conferences and OSHA training classes, its participation in OSHA Telecommunications Task Force efforts, and the dissemination (in English and Spanish) of the NIOSH Alert "Preventing Injuries and Deaths from Falls During Construction and Maintenance of Telecommunication Towers"[12] (NIOSH, 2007e, p. 78).

One concern, however, is that many construction companies are small- to medium-sized enterprises, and the workforce includes a great many non-English speakers. The evidence package does not demonstrate serious attention to the special problems of these stakeholders nor does it outline special efforts to disseminate the information particularly to small businesses.

Outputs

The major outputs of this program include four journal articles, nine conference presentations, one NIOSH publication, 11 FACE reports, and one provisional patent application.[13]

Outcomes

The evidence package documents intermediate outcomes for subgoal 2.3 only, although the many transfer activities documented in a previous section suggest likely intermediate outcomes in the near future. The OSHA Telecommunications Task Force issued a compliance directive,[14] which contains procedures for the inspection of tower construction sites and procedures for tower construction. OSHA now requires 100 percent fall protection, meaning protection for the entire time a worker is off the ground. A revision of that directive removed an important restriction, namely that an erector can now ride a hoist no matter the height off

[12]NIOSH. 2001. NIOSH Alert: Preventing injuries and deaths from falls during construction and maintenance of telecommunication towers. HHS (NIOSH) Publication No. 2001-156.

[13]The provisional patent application was filed in July 2006 for a harness accessory, which automatically supports a wearer in a sitting position with the knees elevated at a position at or above the hips after a fall (CDC Ref. No. I-002-06) (NIOSH, 2007f, p. A-10).

[14]CPL 02-01-029-CPL 2-1.29—Interim Inspection Procedures During Telecommunication Tower Construction Activities.

the ground (a previous directive limited this to 200 feet above the ground). These directives were based on TI Research Program findings (NIOSH, 2007e, p. 79).

NIOSH analyses—including FACE investigations—were used as the basis for the North Carolina Telecommunications Tower Standard, which outlines proper safety procedures to be used during tower construction and maintenance. Other uses of TI Research Program investigations in telecommunications tower-related concerns led to an OSHA safety checklist and a 3-day "train-the-trainer" program for OSHA compliance officers, contractors, component manufacturers, and tower owners and erectors. NATE used the TI Research Program findings as part of the evidence base for a comprehensive safety and health manual as well as a recommended best-practices site safety manual. Other intermediate outcomes include providing the Federal Aviation Administration (FAA) with information it found useful when retrofitting a damaged antenna in the Bahamas and providing the United Kingdom with safety information pertaining to working safely at heights (NIOSH, 2007e, p. 79).

The TI Research Program does not report effects on end outcomes other than to note that there is no evidence that the number of tower erector fatalities has decreased, likely because there is not enough information on the number of erectors being built or repaired to allow for calculation of the rate of fatalities from falls in the industry. The number of telecommunications towers is likely increasing, so it is possible that with better surveillance a decrease in the rate of fatal falls could be documented.

Discussion

The TI Research Program's research on falls from elevations addresses an important source of workplace fatalities, especially in the construction industry. The projects include research on different approaches to reducing the risk of falls from roofs, the development of better fall-arrest harnesses, and an effort to reduce the risk of falls in high-risk industries (telecommunications tower maintenance and construction). The projects appear to have the promise of developing better prevention methods and are focused on important prevention issues relevant to the problems being addressed. The research results have been published and disseminated, and the research group has made productive efforts to work with manufacturers, industry and labor representatives, and other stakeholders in order to develop, carry out, and disseminate these results.

The committee identified some weaknesses with this part of the TI Research Program. There appeared to be no research focusing on falls from the same level, a major source of workplace injuries, and the coordination of this research with NIOSH's Construction Safety Research Program is not clearly evident, although similar research is being conducted in that program. Overall, the program might benefit from better coordination with other programs—including construction, mining, and agriculture—and from expansion to include traumatic injuries due to falls from the same elevation.

GOAL 3: REDUCE INJURIES AND FATALITIES DUE TO WORKPLACE VIOLENCE

There are no subgoals for the TI Research Program's goal on workplace violence. Historically, the program's intramural work has focused on Type 1 violence,[15] or criminal intent, where the perpetrator of the violence enters the business with the intent to commit a crime and has no relationship to a business or its employees. The extramural program targets all types of workplace violence and is intended to complement the intramural program (NIOSH, 2007g, p. 84).

Planning and Production Inputs

Quantitative inputs into the TI Research Program's work on workplace violence were NTOF surveillance data on numbers of fatal and nonfatal occupational injuries resulting from workplace violence, as well as the types of job settings in which workplace violence occurs. The TI Research Program also noted the societal cost of workplace homicide, which was estimated to be approximately $6.5 billion for 1992 to 2001.

The TI Research Program has also received input on workplace violence through national conferences, workshops, and stakeholder meetings. In response to receipt of a $2 million appropriation from Congress in 2002 in support of research on workplace violence research, NIOSH created the Federal Interagency Task Force on Workplace Violence Research and Prevention (NIOSH, 2007g, p. 87).

[15]Workplace violence has been categorized into four types for public health attention (Merchant and Lundell, 2001): Type I is a criminal intent in which the perpetrator has no legitimate relationship to the business or its employees, and is usually committing a crime in conjunction with the violence. These crimes include robbery, shoplifting, and loitering. A large portion of Type I violence occurs in the late-night retail industry. Type II is customer or client violence in which a perpetrator has a legitimate relationship with the business and becomes violent while being served by the business. This category includes customers, clients, patients, students, inmates, and other groups for which a business provides services. A large portion of Type II violence occurs in the healthcare industry, in settings such as nursing homes or psychiatric facilities where the victims are often patient caregivers. Police officers, prison staff, airline employees and teachers provide other examples of workers exposed to this violence. Type III is worker-on-worker violence in which the perpetrator is an employee or past employee of the business who attacks another employee. Type IV is personal relationship violence in which the perpetrator does not have a relationship with the business but with the victim.

The task force is comprised of representatives from 20 federal agencies. NIOSH has coordinated and hosted four task force meetings since 2002 at which attendees share information and identify opportunities for collaboration. This has been an important source of input into the TI Research Program's workplace violence research (NIOSH, 2007f, p. A-14). NIOSH also held a series of stakeholder meetings in 2003 to gather input on workplace violence in the healthcare, retail, and security and law enforcement industries and on domestic violence in the workplace (NIOSH, 2007f, p. A-15–A-16).

The total intramural and extramural combined funding for research on workplace violence for the period FY1997-2005 was $20,254,590. Only the working youth program had higher overall funding for this period. With the exception of FY1997, funding was higher each year for extramural than for intramural research on workplace violence research (see Table 1-2). In FY2002, the TI Research Program received a $2 million appropriation from Congress to support research on workplace violence. The appropriation resulted in a near doubling of total funding. Approximately 25 percent of appropriated funds have supported intramural research and 75 percent have supported extramural research (NIOSH, 2007g, p. 83). FTEs ranged from a low of 4.5 in FY1998 to a high of 9.5 in FY2004.

Activities

TI Research Program activities for workplace violence included surveillance, risk factor research, intervention effectiveness research, and transfer. The committee found these activities to be relevant to the challenge of reducing injuries and fatalities due to workplace violence.

Workplace violence research has addressed high-priority needs. For the period 1990-1998, the TI Research Program focused intramural research on Type 1 violence—criminal intent—the most prevalent type of fatal workplace violence. This research reviewed robbery and robbery-related risk factors for late-night retail workers who are at especially high risk for workplace violence. The findings from this research were also helpful in supporting elements of Crime Prevention Through Environmental Design programs and creating recommendations for the prevention of workplace violence in NIOSH documents and OSHA guidelines.

A very important collaborative intramural activity was the TI Research Program's work to address the high-priority issue of a gap in data on nonfatal injuries from workplace violence. Working with the DOL, the TI Research Program funded and inserted modules of questions to capture data on nonfatal injuries into three national surveys between 2002 and 2006 (NIOSH, 2007g, p. 85). The committee sees this as a significant contribution to the knowledge base on workplace violence and supports plans to complete analyses and publications of findings from these surveys. The availability of data on injuries is very relevant to the challenge of reducing injuries from workplace violence as such injuries are often nonfatal.

The extramural research program is intended to be complementary to the intramural research program and focuses on all types of workplace violence. From 1996 to 2003, NIOSH funded 16 extramural grants—9 for risk factor research and 7 for intervention research. This research targeted several worker populations (retail workers, healthcare and social services workers, police officers, long-haul truckers, and domestic violence in the workplace), but focused mostly on the high-priority retail and health and social service worker populations. The committee found the TI Research Program's work with the University of California, Los Angeles, which evaluated the effectiveness of a Workplace Violence Prevention Program in 314 Los Angeles retail establishments, to be particularly strong and relevant.

Although not a specific target, vulnerable populations have been well addressed in both the intramural and the extramural research activities, particularly small businesses such as convenience stores and taxi drivers.

Transfer Activities

NIOSH and the TI Research Program engaged in several relevant proactive transfer activities for workplace violence. In 2002, NIOSH formed the Federal Interagency Task Force on Workplace Violence Research and Prevention. The meetings provide an opportunity for representatives of 20 federal agencies who are doing work or have an interest in workplace violence research to share information and form partnerships. NIOSH has also partnered with the Veterans Health Administration to evaluate workplace violence interventions in hospitals (NIOSH, 2007f, p. A-14).

TI Research Program staff has provided technical assistance through the development of and participation in workplace violence task forces, workshops, and conferences. One notable effort was a TI Research Program partnership with the University of Iowa in 2000 to hold a workshop focusing on workplace violence intervention research. The conference resulted in the summary report "Workplace Violence: A Report to the Nation"[16] and peer-reviewed papers (NIOSH, 2007f, p. A-15). The conference summary report led to a $2 million appropriation from Congress to the TI Research Program in FY 2002 for workplace violence research. TI Research Program staff also provided technical assistance to employers, employees, and safety professionals through production of the DVD "Violence on the Job."[17] The DVD provides recommendations for preventing work-related

[16]Available at http://www.public-health.uiowa.edu/Iprc/NATION.PDF.

[17]NIOSH. 2004. Violence on the job (DVD). HHS (NIOSH) Publication No. 2004-100d.

homicides and assaults. More than 30,000 copies were distributed (NIOSH, 2007f, p. A-13; NIOSH, 2007g, p. 85).

Several of the TI Research Program's publications have been distributed to federal agencies, such as OSHA, the Office of Personnel Management, and the Federal Bureau of Investigation to assist them with the development of their own workplace violence recommendations (NIOSH, 2007g, p. 85), and thousands of copies of NIOSH publications on workplace violence have been disseminated. TI Research Program staff also responded to media requests for information about workplace violence information (NIOSH, 2007g, p. 85).

Outputs

The TI Research Program has a wide range of outputs in the area of workplace violence including intramural and extramural research publications, workplace violence prevention recommendations, staff interviews with media about workplace violence prevention, presentations at scientific conferences, stakeholder meetings, and national conferences. NIOSH and the TI Research Program have also sponsored or co-sponsored a number of workshops, conferences, and meetings on workplace violence.

Several publications, such as the NIOSH Current Intelligence Bulletin "Violence in the Workplace: Risk Factors and Prevention Strategies"[18] were groundbreaking, because they were among the first to highlight the problem of workplace violence on the national level (NIOSH, 2007g, p. 85). Several other publications addressed the high-risk population of retail workers. As stated previously, there has been a lack of surveillance of nonfatal injuries from workplace violence and the TI Research Program's development of modules for use in three national surveys during 2002 to 2006 to collect nonfatal workplace violence injury data has helped to generate important new knowledge. Several outputs (such as the formation of the Federal Interagency Task Force on Workplace Violence Research and Prevention, research collaborations with universities, stakeholder meetings, and TI Research Program presentations at conferences) also build upon NIOSH's institutional knowledge base and facilitate effective cross-agency and internal-external collaborations.

Outcomes

The evidence package documents several intermediate outcomes, including substantially increased attention to workplace violence by researchers since the late

[18]Jenkins, E. L. 1996. Current Intelligence Bulletin 57, Violence in the workplace: Risk factors and prevention strategies. HHS (NIOSH) Publication No. 96-100.

1980s (NIOSH, 2007g, p. 87). In addition, OSHA appears to have been influenced by TI Research Program outputs. OSHA cited the NIOSH Current Intelligence Bulletin "Violence in the Workplace: Risk Factors and Prevention Strategies"[19] in its recommendations for taxi and delivery drivers as well as its guidelines for healthcare and community service workers. OSHA also cited TI Research Program surveillance and convenience store research in its recommendations for the prevention of workplace violence in late-night retail establishments (NIOSH, 2007g, p. 88). Although it has not been validated, these OSHA recommendations and guidelines are likely to have influenced workplace violence prevention policies in a number of states. TI Research Program staff also contributed to the development of a guide on workplace violence prevention for federal government agency planners by the Office of Personnel and Management (OPM, 1998), as well as guidelines for workplace violence prevention and a response by the American Society for Industrial Security (ASIS, 2006).

As a possible end outcome of its work, the TI Research Program cites a significant decline in workplace homicides for 1993 to 2004. During that time, workplace homicides fell from 1,074 to 559, or almost 48 percent (NIOSH, 2007g, p. 89). The greatest decline was in the retail industry where the number of homicides fell almost 70 percent. A recent publication by NIOSH researchers (Hendricks et al., 2007) expands on this in reporting a significant decline in the rates of occupational homicide of approximately 6 percent per year between 1993 and 2002. This is slightly greater than the decline in rates of all U.S. homicides (5 percent per year) over the same period. The reduction in workplace homicides and the reduction in homicides overall are inextricably linked, and although the TI Research Program may have played a role in the reduction of workplace homicides, the degree of reduction that can be attributed to the TI Research Program's workplace violence efforts versus the decline in homicides overall is unknown.

Discussion

The workplace violence prevention program appropriately focused on selected environments, collaborated with research partners, and achieved notable outcomes that have resulted in guidelines, policies, and progress toward reducing the burden of injuries and deaths due to workplace violence. Extramural research activities dominated after the 2002 federal earmark (75 percent of funds then went to outside researchers). However, there appears to be a good balance of intramural and extramural activities targeting different work environments and different types of workplace violence. Through research and transfer activities, including the dis-

[19]See footnote 18 for reference.

semination of publications on workplace violence, NIOSH and the TI Research Program have helped make workplace violence a public health research priority.

While it is difficult to draw direct inferences, selected occupational subgroups have experienced recent declines in workplace homicides. Replicating this progress for, and identifying technologies that address, other work environments; broadening the input from federal agencies; and continued transfer of research into practice, such as the work done in Los Angeles to test workplace violence interventions in convenience stores, will be vital to continued progress toward reducing the burden of workplace violence and death.

GOAL 4: REDUCE INJURIES AND FATALITIES DUE TO MACHINES

The TI Research Program developed three subgoals for research aimed at reducing injuries and fatalities due to machines:

4.1. Reduce injuries and deaths caused by tractor rollovers by increasing the availability and use of effective rollover protective structures (ROPS)
4.2. Reduce worker injuries and deaths caused by paper balers
4.3. Reduce injuries and deaths caused by machines through the conduct of fatality investigations and the dissemination of prevention strategies

Inputs

More than half (991 out of 1,894) of the agricultural worker deaths from tractor-related events that occurred between 1992 and 2001 were due to tractor overturns. One means of preventing deaths from tractor rollovers is the use of tractor ROPS. Newly manufactured tractors are built with ROPS and older models can be retrofitted. As of 2001, however, about 50 percent of tractors used on farms in the United States did not have ROPS. The TI Research Program's work for subgoal 4.1 addresses barriers to the use of ROPS on tractors including cost and interference with low-clearance farming (which is needed in livestock, dairy, and fruit or nut tree farming). The TI Research Program's work also addressed the problem of outdated anthropometry data for the agricultural populations that operate tractors (NIOSH, 2007h, pp. 97-98).

From 1986 to 2002, two-thirds of the 43 U.S. workers who suffered fatal injuries while operating recycling industry balers were injured using horizontal balers that were baling paper and cardboard (NIOSH, 2007h, p. 104). Workers are at risk for this type of injury when attempting to remove jammed materials from balers that are not properly de-energized or safeguarded. An OSHA control of hazardous energy (or "lockout-tagout") standard (29 CFR 1910.147) establishes requirements for the control of unexpected energization or start-up of machines or equipment, and requires that procedures be developed that either prevent (lock) or warn workers (tag) of possible machine energization or energy release. However, lockout-tagout is not automatic and can be bypassed or forgotten. Subgoal 4.2 attempts to address this problem.

The choice of FACE investigations (subgoal 4.3) as a focus reflects the inadequate information in general surveillance systems identifying individual machines or types of machines associated with worker injuries. Even when injury and fatality data are available, information on the circumstances that led up to the incident, to identify risk factors and intervention options, is often missing. The TI Research Program's FACE program for machines allows for the collection of detailed information on the types of machine injuries that are occurring, the identification of risk factors, and recommendations for prevention (NIOSH, 2007h, p. 107).

The total intramural and extramural combined funding for the machines goal between FY1997 and FY2005 was $15,825,168 ($11,179,879 intramural and $4,645,289 extramural). Program funding had increased over time, especially for the extramural program. At its lowest in FY1997 the combined funding totaled $1,247,075, and at its peak in FY2005 funding reached $2,512,446. FTEs ranged from a low of 9.6 in FY1998 to a high of 14.8 in FY2002 (see Table 1-2).

The committee found all three subgoals appropriate to the overarching challenge of reducing machine-related injuries. The subgoals are narrow and focused on problem areas that the TI Research Program is capable of addressing with its funding and staff resources.

Activities

The TI Research Program has been conducting surveillance of farm worker injuries and fatalities and ROPS use on farms since 1994 when the Traumatic Injury Surveillance of Farmers (TISF) was developed. TISF was discontinued in 1997, but the TI Research Program has continued surveillance activities through the OISPA project (NIOSH, 2007h, pp. 98-99).

A major activity for subgoal 4.1 was the design of cost-effective ROPS (CROPS) made from readily available commercial parts for use on different tractor models and auto-deploying ROPS (AutoROPS) for use in low-clearance farming. Both of these activities addressed important barriers to the use of ROPS on tractors. CROPS are a lower-cost alternative for consumers and thus could help address the cost-of-retrofitting barrier. In 2002, the TI Research Program built, designed, and tested CROPS that passed the Society of Automotive Engineers' (SAE) SAE-J2194 industry ROPS standard (NIOSH, 2007h, p. 100). Building on this effort, the TI

Research Program and a ROPS manufacturer completed CROPS designs for five additional tractor models (NIOSH, 2007h, p. 100). In 1999, the TI Research Program developed AutoROPS as well as an overturn sensor that would monitor tractor operating conditions and deploy AutoROPS. The TI Research Program conducted laboratory and field tests of both devices. The first overturn test, conducted in 2000, showed the AutoROPS to be effective. The TI Research Program also used TISF data to conduct assessments of the cost-effectiveness of retrofitting tractors with ROPS, the results of which showed that retrofitting saves lives and money. One retrofitting assessment that included the cost of nonfatal injuries found that retrofitting tractors would save almost $490,000 per averted injury (NIOSH, 2007h, pp. 99-100).

The TI Research Program's work in addressing barriers to the use of ROPS is to be commended. On the manufacturing end, the TI Research Program has done excellent work to engage manufacturing partners in the ROPS development process. However, many decades-old tractors exist and the retrofitting of older tractors has occurred at a slower than optimal pace. Sometimes the barriers to retrofitting are not related to cost. For example, Hallman (2005) noted a nonfinancial "hassle factor" as a significant obstacle to farmers' willingness to retrofit. As discussed later in this report (see Chapter 3), the TI Research Program could, in the future, devote more resources toward research on how to facilitate faster adoption of the ROPS technology among farm workers.[20]

In response to the problem of outdated anthropometry data for tractor users, the TI Research Program—using body scanning technology—collected three-dimensional body measurements of 100 farmers and farm workers. The program found the vertical clearance suggestion in the current SAE ROPS standard to be about 12 percent too short. The TI Research Program presented its findings to the SAE-J2194 Standard Committee in 2004, but there has not yet been a revision of the standard (NIOSH, 2007h, pp. 98-100).

The major activity for subgoal 4.2 was the design and evaluation of a prototype system to automatically detect a jam in a recycling baler and safeguard workers by powering off. The system, called JamAlert, allows a jam to be cleared without the threat of unexpected energy release. The TI Research Program tested the prototype under a variety of operational conditions. Activities were collaborative; the TI Re-

[20]Although not indicated in the evidence package, the committee learned that, in 2006, NIOSH provided funding to several university-based agricultural safety and health research centers to explore techniques to promote safe use of tractors. Under the initiative, the centers are studying some of the barriers to the use of ROPS on tractors, such as financial incentives to retrofit tractors with ROPS and the impact of changes to standards, regulations, and technology, and their effect on future ROPS availability. Several of the centers are also testing community-based social marketing techniques to improve safe use of tractors (NIOSH, 2006).

search Program worked with members of the ANSI Z245.5 Baler Safety Committee on design of the JamAlert prototype to ensure that it met or exceeded standards. It also consulted with equipment users, builders, and safety device manufacturers throughout the research process (NIOSH, 2007h, pp. 104-105). The committee found this activity to be highly relevant to address the bypassability limitation of the OSHA lockout-tagout standard requirements for machines.

The major activity for subgoal 4.3 has been fatality investigations and dissemination of recommendations. The TI Research Program conducted 45 machine-related fatality investigations from 1994 to 2006, collecting information on events that occurred prior to, during, and after fatal incidents in order to better understand contributing factors and develop recommendations for preventing similar incidents. More than 500 machine-related fatality investigations were conducted through the state-based FACE Program (NIOSH, 2008). Results of fatality investigations are put into narrative reports and made available on the NIOSH Web site (NIOSH, 2007h, p. 107). The TI Research Program also disseminated thousands of copies of NIOSH publications on the prevention of injuries from different types of machines (see following section). Although some types of machine-related injuries will be more relevant than others, the committee is pleased to see that the TI Research Program is looking at preventing diverse types of injuries beyond ROPS and paper balers. The committee supports the TI Research Program's plans to include machine-related fatalities under the FACE Program in the future.

Transfer Activities

Transfer activity for subgoal 4.1 involved work with an equipment manufacturer to develop new AutoROPS for the manufacturer's zero-turn commercial mower. This activity was relevant, and working with industry partners gives them a stake in consumer use of this important safety measure. The TI Research Program petitioned the American Society of Agricultural and Biological Engineers (ASABE) to begin work on a performance standard[21] for the AutoROPS as well. The standard is in draft form and has undergone a first review by ASABE (NIOSH, 2007h, p. 101). The JamAlert system for paper balers (subgoal 4.2) was not ready for transfer to the workplace and the TI Research Program is working with an industry partner to field test and commercialize this technology (NIOSH, 2007h, p. 105). In addition to FACE investigation reports (which are disseminated to target audiences based on the topic area), the TI Research Program disseminated thousands of copies of NIOSH documents (Alerts, Hazard Identification [ID] Bulletins, Workplace Solutions) on the prevention of injury from several different types of machines

[21]ASABE-X599, Standardized Deployment Performance of an Automatic Telescoping ROPS for Agricultural Equipment.

(including cranes, paper balers, forklifts, and excavators or backhoe loaders) for subgoal 4.3. Dissemination was often done in collaboration with partners, facilitating broader transfer.

Outputs

The TI Research Program authored 15 peer-reviewed articles as well as 19 conference presentations on subjects ranging from surveillance data, anthropometry of the farm worker population, economic analyses, and engineering analyses of performance and effectiveness of ROPS, AutoROPS, and CROPS. One manuscript on ROPS use and cost of retrofitting (Myers and Snyder, 1995) was referenced in at least 25 peer-reviewed journal articles. Economic analyses of ROPS retrofitting (Myers and Snyder, 1995; Pana-Cryan and Myers, 2000, 2002) were referenced in a minimum of 32 peer-reviewed journal articles based on a search of a series of manuscripts (NIOSH, 2007h, p. 101). The major output for subgoal 4.2 was the prototype JamAlert, for which the TI Research Program has filed an employee invention report and U.S. patent application. A paper on the laboratory testing that led to JamAlert design recommendations was presented at a 2005 American Society of Mechanical Engineers meeting. Another output was the NIOSH Alert "Preventing Deaths and Injuries While Compacting or Baling Refuse Material,"[22] which provides recommendations for employers on prevention of injuries from working near refuse compacting and baling equipment (NIOSH, 2007h, p. 105). In terms of new technology, ROPS and JamAlert can be best characterized as the application of existing technology and do not appear to be particularly innovative. The committee found these outputs to be relevant to small businesses, and technologies appear to be "user-friendly," although noncompliance may be a problem. Outputs for subgoal 4.3 were 45 investigative summary reports and 10 NIOSH documents addressing a variety of machine-related hazards.

Outcomes

The TI Research Program reports intermediate outcomes for each of the three subgoals for machines. Each of these intermediate outcomes contributes to the desired end outcome of a reduction in machine-related injuries. For subgoal 4.1, Colorado State University used TISF data to help target engineering research on the ability of pre-ROPS tractors to withstand forces of tractor overturn if ROPS were mounted on them and to identify the most common tractors by manufacturer and

[22]NIOSH. 2003. Preventing deaths and injuries while compacting or baling refuse material. HHS (NIOSH) Publication No. 2003-124.

model for ROPS retrofit evaluations (NIOSH, 2007h, p. 101). A strong intermediate outcome for subgoal 4.2 was the contribution of the TI Research Program's work on paper balers to revision of the ANSI Z245.5 Baling Equipment Safety standard. The evidence package reports that since the revision, baler manufacturers are providing purchasers of new balers with safety equipment and safety instructions that meet revised standard requirements. In addition, because hazards for which means of control are generally known are included under the General Duty Clause of the OSH Act,[23] manufacturers should be meeting the standard (NIOSH, 2007h, p. 105). Events following the release of a number of NIOSH publications for subgoal 4.3 are indicative of a positive stakeholder response. One example is that, after the dissemination of the Hazard ID Bulletin "Ignition Hazard from Drilling into Sealed Frames of Agricultural Equipment,"[24] the plow manufacturer stopped using machine shop metal scrap for filling its equipment frames (because it was igniting) and began using clean stainless steel punch-out scrap. A number of publications, including a nationwide Future Farmers of America publication, summarized the hazard after the bulletin was released as well (NIOSH, 2007h, p. 111).

The evidence package notes an end outcome for subgoal 4.3 of a steady decline from 1992 to 2005 in work-related fatalities caused by machines, plant and industrial powered vehicles, or tractors (16 percent decrease in number of deaths; 30 percent decrease in fatality rates) (NIOSH, 2007h, p. 111). Distribution of safety recommendations from fatality investigations and NIOSH publications to reduce machine-related injuries likely played some role in this decline.

Discussion

Machines are a common source of occupational injury and the TI Research Program is to be commended for its attention to this area. The committee found all subgoals for machines to be in line with the National Occupational Research Agenda (NORA I).

Activities for tractor rollovers and paper balers were relevant, focusing on the development of new technologies (ROPS and JamAlert) to prevent worker injuries. The TI Research Program engaged in mutually beneficial collaborations to develop these technologies. In the future, the TI Research Program might consider extending its work on ROPS to technology adoption research to facilitate faster

[23]Part (a) of the OSHA General Duty clause reads: (a) Each employer: (1) shall furnish to each of his employees employment and a place of employment which are free from recognized hazards that are causing or likely to cause death or serious physical harm to his employees; (2) shall comply with occupational safety and health standards promulgated under this Act (DOL, 2008).

[24]NIOSH. 1998. Ignition hazard from drilling into sealed frames of agricultural equipment. HHS (NIOSH) Publication No. 98-146.

adoption of the ROPS technology by farm workers. The machines FACE Program acknowledges the diversity of machines and machine injuries in the workplace, and the important role that FACE investigations play in NIOSH contributions. The committee found the FACE Program to be the shining star of the research program for machines.

In addition to ROPS and JamAlert, FACE investigations and NIOSH publications appeared to impact intermediate outcomes. Particularly strong was the contribution of the TI Research Program's work on paper balers to revision of the ANSI Z245.5 Baling Equipment Safety standard. Several FACE investigation outputs appeared to be well received by stakeholders. Dissemination of these outputs by the TI Research Program likely contributed to declines in numbers and rates of machine-related fatalities between 1992 and 2005.

GOAL 5: REDUCE ACUTE BACK INJURY

NIOSH developed two subgoals for research aimed at reducing acute back injury:

5.1. Reduce acute injuries caused by patient handling
5.2. Evaluate interventions used to prevent acute injuries caused by material handling

Inputs

Planning inputs for subgoal 5.1 were data on the causes, costs, incidence, and prevalence of back injury due to patient lifting. Incidence and prevalence data show that nurses and other caregivers are at high risk for back injury from patient handling. The growth in demand for nurses coupled with their high injury rate, increases in the weight of patients, and the number of older workers in need of assistance with daily tasks were cited by the TI Research Program as causes for concern that back injury may become more of a problem for nurses and other caregivers over time (NIOSH, 2007i, p. 114). Evidence-based training of nurses and caregivers on best practices for patient handling is lacking as well. An outdated curriculum that promotes unsafe patient handling practices continues to be used (NIOSH, 2007i, p. 114).

The only input cited by the TI Research Program for subgoal 5.2 was a need to increase the science on the effectiveness of back belts for preventing back injuries among materials handlers. Findings from prior studies of this issue showed conflicting results and some studies had design limitations.

The total intramural and extramural combined funding for back injury research in the TI Research Program between FY1997 and FY2005 was $10,257,669

($7,329,287 intramural and $2,928,382 extramural). In terms of funding, this was the smallest of the TI Research Program's eight goals for this period. Funding for both intramural and extramural research has generally decreased over the years. Combined funding was at its lowest in FY2004, when there was zero funding for extramural back injury research. FTEs have decreased over time as well, from a high of 15.7 in FY1998 to a low of 3.6 in FY2005 (see Table 1-2). The committee learned from TI Research Program staff that it plans to move work on acute back injury out of its program and into the musculoskeletal program.[25] The decreases in funding may be reflective of that shift.

Activities

The major activity for subgoal 5.1 was an expansion of extramural work done in the early 1990s showing that a safe lifting program using mechanical lifts can be effective in nursing homes. To address the limitations of a small sample size and the relatively short follow-up period of 1 year in the 1992 study, the TI Research Program conducted a large nursing home intervention trial over a 6-year period. The TI Research Program collaborated with a St. Louis, Missouri, health system, which provided the study population. The study tested the effectiveness of a "best practices" in safe lifting program, which was found to significantly reduce injuries for full- and part-time nurses (NIOSH, 2007i, p. 115). The committee found this work to be relevant in that it addresses a knowledge gap. The TI Research Program also collaborated with a number of partners. In addition to the St. Louis, Missouri, health system, the TI Research Program worked with equipment manufacturers to test and evaluate the lifting equipment that was used in the intervention study.

The primary activity for subgoal 5.2 was a large prospective cohort study conducted from 1996 to 1998 to evaluate the effectiveness of industrial back belts at preventing initial and recurrent low back injuries. The TI Research Program collaborated with Wal-Mart to enroll materials handlers with the highest lifting exposures from 160 Wal-Mart stores. Results from the study showed that neither frequent back belt use nor a store policy requiring the use of back belts reduced worker lower back injury or pain, and were suggestive to the TI Research Program that back belts could not be recommended for general use as protective technology in the workplace (NIOSH, 2007i, p. 123). The TI Research Program also conducted two laboratory evaluations of the same back belts used in the cohort study. Significant findings from the laboratory studies were that use of back belts decreases mean oxygen consumption, forward spine bending, and velocity of forward and

[25]Personal communication from N. Stout.

back spine bending. The committee sees these activities as relevant, because many work-related back injuries are caused by materials handling, and consistent with the subgoal 5.2 objective. However, in addition to industrial back belts, other interventions to prevent injuries due to materials handling could have been explored. The work in this area was collaborative; in addition to Wal-Mart, the TI Research Program worked with two universities and a survey research organization to administer telephone surveys of study participants and manage data (NIOSH, 2007i, pp. 123-124).

Transfer Activities

The TI Research Program has engaged in a variety of transfer activities for subgoal 5.1. Findings from the work on safe patient lifting were published in the NIOSH document "Safe Lifting and Movement of Nursing Home Residents"[26] and disseminated to thousands of nursing homes in the United States. The TI Research Program staff contributed to the work of a DOL-OSHA committee to synthesize evidence on safe patient lifting programs from 1998 to 2002. This work resulted in a publication of OSHA guidelines for nursing homes. To address the problem of outdated training curriculum for nurses, the TI Research Program has been working with the National Council Licensure Examination Board to update safe patient lifting curricula and include the latest research findings in nursing licensure exams (NIOSH, 2007i, p. 116). The committee supports the TI Research Program's plans to pursue additional work in this area, such as the evaluation of training curricula in nursing schools (NIOSH, 2007i, p. 119). The TI Research Program might also evaluate the impact of state legislation to promote safe patient handling.

There did not appear to be much proactive transfer activity for subgoal 5.2. The Wal-Mart study was published in the *Journal of the American Medical Association* (*JAMA*) (Wassell et al., 2000). The laboratory evaluation studies were published in *Applied Ergonomics* (Bobick et al., 2001) and *Spine* (Giorcelli et al., 2001). NIOSH maintains a summary of findings on back belt research on its Web site.

Outputs

For subgoal 5.1, outputs included peer-reviewed journal articles, NIOSH publications, conference proceedings and presentations, and chapters in nursing student textbooks. Outputs build internal and external knowledge about back injury due to patient handling and support research and education capacity. For example, the

[26]Collins et al., 2006. Safe lifting and movement of nursing home residents. HHS (NIOSH) Publication No. 2006-117.

"Safe Lifting and Movement of Nursing Home Residents"[27] presents a business case to nursing home workers interested in instituting a safe patient lifting program. The work to update nursing training curricula helps to ensure that nurses are provided an evidence-based education on patient lifting. The committee finds that outputs could have better accommodated Spanish-speaking workers who make up increasing numbers of nursing home employees.

The outputs for subgoal 5.2 were peer-reviewed journal articles from the Wal-Mart and laboratory studies on industrial back belts (NIOSH, 2007i, p. 124). In the cohort study, the TI Research Program found that neither frequent back belt use nor a store policy that employees must wear back belts was associated with a reduced incidence of back injury claims or low back pain (NIOSH, 2007i, p. 123). This output adds to the evidence that back belts may do little to prevent back injury caused by materials handling.

The committee noted a reduction in outputs for the TI Research Program back injury research over time that may reflect NIOSH's plans to combine its acute back injury research with chronic, musculoskeletal back injury research. The committee believes that this is in NIOSH's best interest.

Outcomes

TI Research Program work on back injuries has resulted in workplace changes through the passage of legislation. Patient handling research was used in support of the passage of safe patient handling legislation in six states from 2005 to 2006 (NIOSH, 2007i, pp. 116-117).[28] Legislation addressed safe patient handling policies in hospitals and nursing homes, provision of lifting equipment to hospitals, education and training of employees, and safe patient handling research, among other things. Six additional states have proposed similar legislation based on TI Research Program work. The American Nurses Association is using TI Research Program findings on patient handling in its "Handle with Care" program, which is a national effort to reduce musculoskeletal injuries among nurses due to patient handling (NIOSH, 2007i, p. 117).

The TI Research Program cites a decrease in incidence rate for sprains and strains in hospitals and nursing homes from 1992 to 2005, more specifically in cases where lifting patients was the source of the injury, as evidence of progress toward an end outcome (NIOSH, 2007i, pp. 117-118). A direct connection cannot be made between the TI Research Program's work and this decline. However, the

[27] See footnote 26 for reference.

[28] Legislation included Texas Senate Bill 1525; Washington House Bill 1672; Hawaii House Concurrent Resolution No. 16; Rhode Island House 7386 and Senate 2760; Ohio House Bill 67; and New York companions Bills A07641 and S04929.

committee agrees that the TI Research Program has provided much of the knowledge base that would allow workplaces to make changes that result in reductions in musculoskeletal injuries from lifting patients.

According to the evidence package, publication of the Wal-Mart study in *JAMA* had an impact on the use of back belts in a number of large retail outlets. Many stores withdrew store policies requiring the use of back belts and some stores began to use other back injury prevention measures (NIOSH, 2007i, p. 124). Because the TI Research Program's work in materials handling led to the knowledge that back belts were ineffective, the committee did not expect an end outcome for this subgoal.

Discussion

The TI Research Program's work on back injuries is very relevant. Back injuries continue to be among the most serious occupational problems in terms of injury and lost time. Both subgoals are aligned with NORA I priorities.

The major activity for subgoal 5.1 addressed a research gap by evaluating a safe patient lifting best-practices intervention. Activities for subgoal 5.2 were narrowly focused on evaluating back belts and patient lifting devices. A well-designed cohort study added to the evidence on the effectiveness of back belts. However, the committee feels the program would have been stronger had other intervention strategies to address manual materials handling also been evaluated.

Outputs for subgoal 5.2 were comprehensive and add to the knowledge about back injury due to patient lifting. Intermediate policy outcomes for this subgoal (legislation in six states) were particularly strong in that they have the potential to result in workplace changes to reduce risk of back injuries associated with patient lifting.

GOAL 6: REDUCE INJURIES AND FATALITIES AMONG WORKERS IN ALASKA

The TI Research Program developed an Alaska goal area after discovering in the late 1980s that the state had a higher worker fatality rate than any other U.S. state. From 1980 to 1989, the annual traumatic occupational fatality rate in Alaska was nearly 35 per 100,000 workers, or almost five times the annual U.S. average of 7 per 100,000 workers (NIOSH, 2007j, p. 126). NIOSH established the Alaska Field Station (AFS) in Anchorage in 1991 with a goal of reducing Alaska's high rate of traumatic occupational fatalities. The work was supported by a 1990 congressional directive for NIOSH to "engage in research, prevention, and outreach activities to reduce these unacceptably high levels of preventable occupational injuries."

The TI Research Program developed three subgoals for research aimed at reducing injuries and fatalities among workers in Alaska:

6.1. Reduce injuries and fatalities in commercial fishing
6.2. Reduce injuries and fatalities in helicopter-logging operations
6.3. Reduce injuries and fatalities in Alaska aviation

Inputs

One of the first efforts of the TI Research Program in Alaska was to establish a comprehensive occupational injury surveillance system. Although the establishment of this system in 1991 predates the period of the committee's assessment, the surveillance work continues and is key to current activities. The Alaska Occupational Injury Surveillance System (AOISS) and the Alaska Trauma Registry (ATR) were the result of partnerships between federal and state agencies. Also in 1991, the TI Research Program formed the Alaska Interagency Working Group (IAWG) for the Prevention of Occupational Injuries to provide a broader understanding of occupational injuries in the state and to enable rapid response to emerging problems. The IAWG is made up of members from state and federal agencies, local entities, industry associations, and the University of Alaska at Anchorage. There are IAWG subgroups for Alaska commercial fishing safety, aviation, and construction. The IAWG has facilitated the TI Research Program's work in Alaska, including the development of AOISS (NIOSH, 2007j, pp. 128-129).

Inputs specific to subgoal 6.1 were data on risk factors for fatal and nonfatal injuries among commercial fishers. Fisherman fatalities were often found to be due to the sinking of vessels and crew members falling overboard, while nonfatal injuries more often occurred on deck during deployment and retrieval of fishing gear. The TI Research Program also wanted to evaluate commercial fishing industry policies that had not yet undergone evaluation, such as the 1988 Commercial Fishing Industry Vessel Safety Act (CFIVSA)[29] (NIOSH, 2007j, p. 127). The TI Research Program pursued subgoal 6.2 based on data showing that, over an 18-month period between January 1992 and June 1993, 6 of 25 helicopters used for logging in southeast Alaska crashed, resulting in 9 deaths and 10 severe injuries. Investigations by the National Transportation Safety Board (NTSB) had shown improper operation and/or vehicle maintenance to be a problem in all of these crashes (NIOSH, 2007j, p. 137).

Alaska is dependent on air transportation for transport of goods and passengers to and from areas not accessible by other means, yet aviation is a high-risk

[29]P.L. 100-424, September 9, 1988, 100th Congress: To provide for the establishment of additional safety requirement for fishing industry vessels, and for other purposes.

industry in Alaska due to the state's unique terrain and weather challenges as well as incomplete radar coverage (NIOSH, 2007j, p. 137). From 1990 to 1999, 401 people died in Alaskan aviation crashes and almost 40 percent of fatalities were of workers on the job. Controlled flight into terrain (CFIT)[30] accounted for an average of five fatal occupational crashes per year in Alaska for this period (NIOSH, 2007j, p. 141). In addition to these data, the TI Research Program's work for subgoal 6.3 is influenced by legislation passed by Congress in 2000 (P.L. 106-69) to reduce occupational aircraft crash fatalities by 50 percent by the end of 2009. This legislation resulted in formation of the Alaska Interagency Safety Initiative to improve air safety through the collaborative efforts of the TI Research Program, the FAA, the National Transportation Safety Board, and the National Weather Service (NIOSH, 2007j, p. 141).

Stakeholder input is a staple to the Alaska program, in part through the IAWG. However, there was no mention of stakeholder input from labor organizations or other worker representatives in the materials reviewed. In addition, the committee would have liked to have seen more evidence of collaborative involvement of Native Alaskan groups included.

The total intramural and extramural combined funding for the Alaska workers program between FY1997 and FY2005 was $14,089,400 ($10,932,226 intramural and $3,157,174 extramural), making it one of the smaller programs for this period in terms of funding (see Table 1-2). Extramural funding decreased each year from 2001 to 2005, and in 2005, no funds were allocated for extramural research. FTEs ranged from a low of 6.6 in FY1999 to a high of 12.8 in FY2002, and have been relatively stable since 2001.

Activities

Activities for subgoal 6.1 were broad (including surveillance, risk factor intervention, and policy evaluation research) and relevant to reducing traumatic injuries among commercial fishermen. The TI Research Program is engaged in continuous collection and analysis of surveillance data from the AOISS and the ATR to identify commercial fishing hazards and track progress in the attainment of program goals. An impressive focused intervention effort was a 2000 project to find and disseminate solutions to fatal and nonfatal injuries that occur on fishing vessel decks. Working with the North Pacific Fishing Vessel Owners Association, the TI Research Program conducted focus groups with crab fishermen on deck safety problems and toured vessels to view problems and discuss potential modifications

[30]CFIT is an accident in which an airworthy aircraft inadvertently flies into terrain, an obstacle, or water. Pilots are generally unaware of danger until it is too late.

(NIOSH, 2007j, p. 129). This work resulted in a crab fishermen deck safety handbook[31] and a 1:10-scale deck safety model for use at industry trade shows. Another important intervention activity was the development of an emergency stop—or e-stop—system by the NIOSH Spokane research lab, which allows fishermen to stop a winch quickly when entangled. The TI Research Program worked with a partner on design, installation, and testing of the system in one vessel. A TI Research Program evaluation of the CFIVSA showed that, since passage of the act, there has been a decline in the number of fatalities among commercial fishermen but not in the number of vessels sunk. This activity resulted in recommendations to modify the CFIVSA approach to focus on the prevention of disasters rather than the use and availability of safety equipment during and after disasters and in the NIOSH document "Commercial Fishing Fatalities and Prevention Strategies in Alaska,"[32] which describes risk factors and prevention strategies for commercial fishing deaths in Alaska (NIOSH, 2007j, pp. 129-130).

The TI Research Program arranged an emergency meeting of the Alaska IAWG in 1993 upon discovering a series of helicopter-logging accidents. Meeting participants (FAA, NTSB, OSHA, the U.S. Coast Guard [USCG], the Alaska Department of Labor [AKDOL], and the Alaska Department of Health and Social Services) developed recommendations to improve helicopter-logging safety and called for industry-wide standards and procedures. In response to the meeting, the FAA and AKDOL inspected all helicopter-logging sites and shut down some because of irregularities (NIOSH, 2007j, p. 137). The TI Research Program's activities for subgoal 6.2 for the evaluation period have been a continuation of the IAWG efforts to generate recommendations for the logging industry and to transfer those recommendations to industry (see following section). For example, between March 1995 and March 1997, the TI Research Program cosponsored three Helicopter-Logging Safety Workshops in Alaska for capacity building. Attendees from government, industry, academia, and insurance developed and refined recommendations to be provided to helicopter-logging industry representatives (NIOSH, 2007j, pp. 137-138). There have been no helicopter-logging crashes in Alaska since 1996. The committee finds work by the TI Research Program and partners to maintain this statistic through continued recommendation development and dissemination to be relevant and important.

Activities for subgoal 6.3 seemed to focus predominately on transfer and are described in the section below. One risk factor effort examined practices and attitudes of Alaskan commercial commuter and air taxi operators and their pilot and

[31]Jensen Maritime Consultants. 2002. Deck safety for crab fishermen. Available at: http://www.jensenmaritime.com/images/stories/crabdeck.pdf (accessed August 6, 2008).

[32]NIOSH. 1997. Commercial fishing fatalities and prevention strategies in Alaska. Current Intelligence Bulletin (CIB) #58. HHS (NIOSH) Publication No. 97-163.

companies' fatal accident rates (NIOSH, 2007j, p. 142). Results showed that pilots at operations with high fatalities had less overall career flight experience but worked more hours per week. Another effort evaluated factors related to crash survivability (e.g., post-crash fire, distance from airport).

Transfer Activities

The committee found all transfer activities for the Alaska program to be relevant and sensitive to stakeholder needs.

Subgoal 6.1 transfer activities involved technical assistance from TI Research Program staff. In 2005, the TI Research Program assisted the USCG with an evaluation of the USCG Dockside Pre-season Boarding Program. Findings suggested the program had likely been effective; from 1999 (when the program was implemented) until 2005, there were three fatalities in the crab fishery, compared to an average of seven deaths per year in years prior to 1999 for which data were captured (NIOSH, 2007j, p. 129). TI Research Program staff also provided expert testimony to the National Research Council (NRC) Ocean Studies Board on safety implications of individual fishing quotas[33] for the halibut-sablefish fishery in 1997 and provided a demonstration of the e-stop system at a large U.S. commercial fishing trade show. The crab fishermen deck safety handbook is posted on several safety Web sites, and more than 4,000 copies were proactively distributed to fishermen in the Pacific Northwest.

Transfer activities for subgoal 6.2 involved the dissemination of recommendations and workshop proceedings. In addition to the recommendations from the 1993 meeting, the IAWG disseminated recommendations from the three Helicopter Logging Safety Workshops (NIOSH, 2007j, p. 138).

The TI Research Program has engaged in a wide range of relevant and proactive transfer activities for subgoal 6.3. These include training and information dissemination at seminars and meetings, dissemination of research articles and recommendations, and staff participation on Alaska safety committees. Among the strongest of these activities were aviation safety seminars and meetings targeting passenger and flight crew audiences. In 2000 and 2001, the TI Research Program organized two free Aviation Passenger Safety Seminars and gave presentations on topics such as aviation safety for passengers, training to learn basics of flying and landing a small aircraft (if pilot becomes incapacitated), and incorporation of safety measures into a company-air carrier contract for transportation of employees. Based on intermediate outcomes for this subgoal (described below), this effort

[33]Individual fishing quotas were implemented as an element of fishery management for the halibut-sablefish fisheries in 1995. They set an allowable catch limit for vessel owners and a time frame in which they have to catch their limit (NIOSH, 2007j).

seems to have been very well received. TI Research Program staff has also presented aviation safety information to pilots and industry—for example, at NIOSH-sponsored Aviation Safety Alliance meetings for pilots and industry in 2003 and 2006. The TI Research Program provided findings from crash survivability research to safety organizations and regulatory agencies so they can develop safety programs (NIOSH, 2007j, p. 142).

Outputs

Outputs for subgoal 6.1 were NIOSH publications, journal articles, recommendations, and a new technology intervention (e-stop). In addition, the TI Research Program and partners organized conferences and workshops on commercial fishing safety. These included two Fishing Industry Safety and Health (FISH) Workshops and three International Fishing Safety and Health (IFISH) Workshops, which were important for bringing together representatives and organizations with a stake in commercial fishing safety (NIOSH, 2007j, p. 130). Proceedings of the FISH and IFISH conferences were published as NIOSH documents.

Outputs for subgoal 6.2 were recommendations from the IAWG and Helicopter Logging Safety Workshops, workshop proceedings, and three publications (an article in *Morbidity and Mortality Weekly Report*,[34] a chapter in the book *Safety and Health in Agriculture, Forestry, and Fisheries*,[35] and the NIOSH report "Helicopter Logging Safety"[36]).

Outputs for subgoal 6.3 included meetings and seminars, recommendations, journal articles, and NIOSH publications (one presenting results of an analysis of survey data called "Survey and Analysis of Air Transportation Safety Among Air Carrier Operators and Pilots in Alaska"[37]).

Outcomes

The evidence package noted intermediate and end outcomes for each of the Alaska program subgoals.

---

[34]CDC. 1994. Risk for traumatic injuries from helicopter crashes during logging operations—Southeastern Alaska. *Morbidity and Mortality Weekly Report* 43(26):472-475.

[35]Langley, R. L., R. McLymore, W. J. Meggs, and G. T. Roberson. 1997. Epidemiology and prevention of helicopter logging injuries. In *Safety and Health in Agriculture, Forestry, and Fisheries*. Government Institutes.

[36]NIOSH. 1998. Helicopter logging safety: Alaska interagency working group for the prevention of occupational injuries. HHS (NIOSH) Publication No. 98-147.

[37]NIOSH. 2007. Survey and analysis of air transportation safety among air carrier operators and pilots in Alaska. HHS (NIOSH) Publication No. 2007-102.

One of the strongest intermediate outcomes for subgoal 6.1 was work by the Alaska USCG to design and implement the Dockside Pre-season Boarding Program, which was based on work done at the TI Research Program–sponsored 1997 FISH Workshop. USCG vessel safety examiners developed the "at-the-dock" boarding program to identify and correct safety hazards known to exist in the Bering Sea crab fisheries.[38] The program involves the examination of vessels for stability and lifesaving equipment—as required by CFIVSA—prior to the opening of the crab fishery. If an examination shows that a vessel is not loaded properly or if there is a lack of lifesaving equipment, the vessel is not able to leave port. Another strong intermediate outcome was use of the NIOSH document "Commercial Fishing Fatalities and Prevention Strategies in Alaska"[39] by a USCG task force in 1998 to develop a national plan for fishing vessel safety. In the task force's final report, called "Living to Fish, Dying to Fish,"[40] the USCG adopted 8 of the 11 recommendations from the NIOSH document. The NIOSH document has been used by other organizations as a resource on commercial fishing vessel safety as well. An end outcome noted in the evidence package for subgoal 6.1 is a decline in numbers of deaths among commercial fisherman by 74 percent between 1990 and 2005, and a 51 percent decline in the annual fatality rate during this same period. The overall reduction in fatalities has been driven to a large extent by reductions in fatalities in crab fishery, for which the fatality rate dropped by more than 50 percent comparing rates since 2000 to those of the 1990s (NIOSH, 2007j, p. 132). Although the committee cannot say with certainty that these declines are due to the TI Research Program's work, it agrees that outputs and intermediate outcomes of the TI Research Program and partners, particularly those related to safety for crab fishermen, likely played a role.

An important intermediate outcome for subgoal 6.2 was the adoption of helicopter safety guidelines developed by the 1996 Helicopter Logging Safety Committee in the operational procedures of 27 helicopter logging companies. The evidence package also notes an end outcome for this subgoal. After the meeting of the Alaska IAWG and the helicopter logging site inspections in 2003, no helicopter logging crashes or fatalities occurred until 1996, when one crash took place. No additional crashes have occurred since the 1996 crash despite extensive logging between 1994 and 1999 (NIOSH, 2007j, p. 139). The committee agrees that the rather sharp decline in helicopter logging crashes is likely tied to the TI Research Program's work

[38]These fisheries were chosen based on NIOSH findings identifying the crab fishery as having the highest fatality rate of any fishery in Alaska.

[39]NIOSH. 1997. Current Intelligence Bulletin 58, Commercial fishing fatalities and prevention strategies in Alaska. HHS (NIOSH) Publication No. 97-163.

[40]USCG. 1998. Living to fish, dying to fish. Available at: www.uscg.mil/hq/gm/moa/docs/fvctf.doc (accessed July 23, 2008).

with partners to develop and disseminate recommendations to improve helicopter logging safety (NIOSH, 2007j, p. 138).

TI Research Program outputs for subgoal 6.3 contributed to a number of important educational programs and materials. Information presented by TI Research Program staff at the Aviation Passenger Safety Seminars in 2000 and 2001 was incorporated into the FAA Circle of Safety consumer education program, which aims to educate passengers on their role in aviation safety in Alaska. Information from the seminars and TI Research Program-CFIT research was also incorporated into an FAA informational handout for pilot training. The Alaska Airmen's Association and the FAA used TI Research Program recommendations to develop public service announcements and a crash survival training video. Results from the pilot and operator survey were used by the FAA Flight Standards Division to review intervention programs and by the nonprofit Medallion Foundation to focus pilot training (NIOSH, 2007j, p. 144). End outcomes noted by the TI Research Program for subgoal 6.3 are declines in aviation crashes and pilot fatalities, comparing data from the 1990s to those since the start of the Alaska Interagency Safety Initiative (2000-2005); occupational fatalities in aviation crashes decreased 44 percent and the pilot fatality rate decreased 53 percent. In addition, the average number of fatal occupational accidents due to CFIT declined by 60 percent, from an average of five fatalities per year from 1990 to 1999 to an average of two fatalities per year from 2000 to 2005. The TI Research Program acknowledges that these changes came about because of work done in concert with other agencies, organizations, and industry (NIOSH, 2007j, p. 144). It is impossible to estimate the degree to which the TI Research Program's work contributed to these reductions. However, increased awareness of aviation safety through the TI Research Program and partner consensus building, program outputs, and intermediate outcomes likely played some role.

Discussion

The committee found all elements of the Alaska program to be based on relevant statistics and designed to address important risks in various Alaska work sites. Stakeholder and worker participation seems to be a staple of the Alaska program, although there was no mention of stakeholder input from labor organizations or other worker representatives. There was a surprising lack of collaborative involvement of Native Alaskan groups.

Activities have centered around an effective model for targeted, applied research using a public health approach at the state level that develops and uses good surveillance data, targets the most significant problems, builds partnerships, develops diverse solutions (technical, policy, training, etc.), implements promising interventions, evaluates impact, and broadly disseminates results. A significant factor in the success of the program, in addition to those listed above, is the investment of sufficient resources, particularly qualified staff. The committee notes that it would be useful to see more interaction among the Alaska program and other researchers in NIOSH so that those other researchers can learn from the AFS approach to applied research and the Alaska team can tap into other research skills at NIOSH (as well as extramural investigators). There has been good transfer through technical assistance and dissemination of documents and recommendations, with clear relevance to reducing traumatic injury in the fishing, logging, and aviation industries. In addition to continued work to target workers in Alaska, the committee is of the understanding that the AFS staff will help address problems with similar industries in the Pacific Northwest and other regions of the United States.

The TI Research Program has documented impressive intermediate and end outcomes for each subgoal. The end outcomes—reductions in deaths of commercial fishermen in Alaska, in crashes of helicopters flying in logging operations, and in pilot fatality rates—generally were likely tied to program outputs and intermediate outcomes.

GOAL 7: REDUCE INJURIES AND FATALITIES TO EMERGENCY RESPONDERS

NIOSH developed three subgoals for research aimed at reducing injuries and fatalities to emergency responders:

7.1. Reduce injuries and fatalities to firefighters
7.2. Improve protection for ambulance workers in patient compartments
7.3. Improve protection for emergency workers responding to large-scale disasters and terrorist attacks

Inputs

Congress directed NIOSH to engage in firefighter safety research (subgoal 7.1) with a special appropriation of $2.5 million in 1998 (NIOSH, 2007d, p. 148). The specific language was: "The conference agreement provides increases above the 1997 level for occupational safety and health for the following activities: intramural research at the Morgantown, West Virginia facility; the fire fighter safety initiative; and the national occupational research agenda."[41]

Planning inputs into subgoal 7.2 were findings from fatality investigations showing inadequate protection for firefighters and other EMS (emergency medical services) workers injured in patient compartments during crashes (NIOSH,

[41]Personal communication from N. Stout.

2007d, p. 150). Lap belts in ambulance compartments do not allow EMS workers the mobility needed to access patients and medical equipment. Therefore, the lap belts may not be used. The TI Research Program also noted the problem of EMS equipment often carried unsecured in ambulance patient compartments, which can become projectiles in the event of a crash.

Subgoal 7.3 addresses the need to learn from disasters and terrorist attacks so that response to such occurrences in the future is safer for emergency response workers.

In planning its work on emergency responders, the TI Research Program received input from fire unions, fire departments, state organizations, and ambulance manufacturers. Workshops with these groups were also held to gain input in dissemination efforts. This interaction appears to have been productive.

The total intramural and extramural combined funding for this project for the period FY1997-2005 was $15,780,346 ($12,090,779 intramural and $3,689,567 extramural). In FY1997 this was the smallest of the eight goals, with combined funding of $166,680 (see Table 1-2). However, funding has increased and, since FY1999, this goal has been one of the TI Research Program's larger research programs. FTEs ranged from a low of 2.5 in FY1998 to high of 18.5 in FY2003.

Activities

In response to the congressional appropriation for firefighter safety research, the TI Research Program initiated the Fire Fighter Fatality Investigation and Prevention Program (FFFIPP) in 1998. This program has been the focus of the TI Research Program's research activities on firefighter safety. The committee finds this activity to be highly relevant to better understanding the risk factors for firefighter line-of-duty deaths and the development of interventions. The FFFIPP involves onsite investigations of firefighter line-of-duty deaths to generate reports describing the circumstances that led to the incidents. Recommendations to prevent similar incidents in the future are developed based on report findings. From 1998 to 2006, the TI Research Program conducted 366 investigations of incidents resulting in firefighter deaths (NIOSH, 2007d, p. 148). The TI Research Program's firefighter safety activities have been very collaborative. One strong example is a TI Research Program-United States Fire Administration memorandum of understanding on how "to identify collaborative efforts to improve safety and health and health conditions for firefighters throughout the United States" in order to facilitate the use of TI Research Program products in firefighter safety training (NIOSH, 2007d, p. 149).

Activities for subgoal 7.2 were relevant and addressed subgoal inputs. In addition, many ambulances are based in fire departments, making this work applicable to both firefighters and other EMS workers (NIOSH, 2007d, p. 157). The

TI Research Program conducted laboratory research with a number of partners to evaluate mobile occupant restraint systems[42] using computer modeling and dynamic testing (NIOSH, 2007d, p. 158). The TI Research Program also worked with a contractor on a performance evaluation of energy-absorbing foams, the findings of which will be used to create recommendations for the EMS industry, ambulance manufacturers, and standards-setting bodies. The TI Research Program used a contractor to evaluate crashworthiness of EMS medical equipment carried in ambulance compartments (NIOSH, 2007d, p. 159). Another important aim of the TI Research Program's work on ambulance safety has been to describe and quantify the scope of injuries among EMS workers. Efforts to address this gap in data included an analysis of data from the NHTSA Fatality Analysis Reporting System, which identified 300 fatal ambulance crashes resulting in deaths of 27 EMS workers from 1991 to 2000. The TI Research Program has also been working with a large EMS provider, American Medical Response, to evaluate the provider's employee injury and compensation database to assess nonfatal injuries to EMS workers due to crashes (NIOSH, 2007d, p. 158).

Subgoal 7.3 was initiated in the aftermath of the September 11, 2001, terrorist attacks. One effort in this area was NIOSH and RAND (Research and Development Corporation) individual and group interviews of emergency response workers to gather input into a collaborative study of occupational safety and health management practices in major disaster response. In 2003, NIOSH and RAND also held a workshop of experts in emergency response and occupational safety and health in order to gather expert opinion on training, hazard assessment, surveillance of emergency workers, and other issues (NIOSH, 2007d, p. 166).

Transfer Activities

The TI Research Program investigations of firefighter fatalities (subgoal 7.1) have resulted in a number of educational documents and journal articles that summarize hazards and recommend prevention measures. Dissemination of these products occurs primarily through the TI Research Program-FFFIPP Web site, which has links to all fatality investigations and other TI Research Program publications. The TI Research Program also engages in proactive transfer of products on firefighter safety. For example, there are periodic mass mailings

[42]"Mobile occupant restraints employ worker-worn harnesses that are tethered to the structure of the patient compartment. The tethers are stored on retractor reels, which unwind as the occupant moves away from the seat and wind up as the occupant moves back into the seat. In the event of a crash, the retractor reels automatically lock, providing restraint against crash forces. This arrangement allows EMS workers the mobility needed to access patients while simultaneously providing crash protection" (NIOSH, 2007d, p. 157).

(approximately one per year) of investigation reports to more than 30,000 U.S. fire departments. Firefighter safety recommendations have been disseminated to equipment manufacturers, municipalities for organization and coordination of fire services and building safety requirements, and research organizations for development and enhancement of safety technology (NIOSH, 2007d, pp. 150-151). TI Research Program staff has provided technical assistance through participation on several National Fire Protection Association (NFPA) technical committees developing standards to support firefighter safety and health.[43] The TI Research Program's Personal Protective Technology Program staff also conducted physiological-ergonomic testing of a prototype personal protective equipment firefighter ensemble as part of Project HEROES (Homeland Emergency Response Operational Equipment Systems) (NIOSH, 2007d, p. 152).

For subgoal 7.2, the TI Research Program has given briefings of tests of mobile occupant restraint systems to research partners. The TI Research Program also provided data packages to partners to support additional research and development. According to the evidence package, the TI Research Program is still working with a contractor to finalize and publish ambulance crash test findings (NIOSH, 2007d, p. 161). The committee supports the TI Research Program's continued analysis and publication of crash data and encourages dissemination of findings for use by others for ambulance design improvements. The contractor evaluating the crashworthiness of EMS equipment in ambulance compartments is continuing work to develop procedures to quantify crash performance of mounting systems (NIOSH, 2007d, p. 159).

A major transfer activity for subgoal 7.3 was the dissemination of the RAND-TI Research Program publication "Protecting Emergency Responders, Volume 3: Safety Management in Disaster and Terrorism Response,"[44] which was released June 2004. The report was launched by press release and presented at a congressional briefing. It was then disseminated by mail to state emergency management and public safety officials, public health and federal agencies, state OSHA programs, and others. The report was also distributed at conferences, trade shows, and exhibits and is available online (NIOSH, 2007d, pp. 167-168). Other transfer work for subgoal 7.3 was TI Research Program staff assistance with injury and illness surveillance following natural and human-caused disasters. In the aftermath of the September

[43]These standards were NFPA 1500 (Standard on Fire Department Occupational Safety and Health Programs), NFPA 1981 (Standard on Open-Circuit Self-Contained Breathing Apparatus for Fire and Emergency Services), NFPA 1982 (Standard on Personal Alert Safety Systems [PASS]), and NFPA 1989 (Standard on Breathing Air Quality for Fire and Emergency Services).

[44]NIOSH. 2004. Protecting emergency responders, volume 3: Safety management in disaster and terrorism response. HHS (NIOSH) Publication No. 2004-144, RAND Publication No. MG-170.

11, 2001, attacks, for example, TI Research Program staff facilitated dissemination of rescue worker injury and illness data to other NIOSH divisions for analysis. TI Research Program staff also designed data collection tools for capturing responder injuries and illnesses as a part of a disaster response worker registry, and provided modular data collection tools that could be used or adapted to monitor injury and illness of workers following the 2004 tsunami in Southeast Asia and Hurricane Katrina in 2005 (NIOSH, 2007d, pp. 165-166).

Outputs

In its work on firefighters, the TI Research Program has produced a large number of fatality investigation reports on individual firefighter deaths and has presented its findings at conferences and meetings as well as in trade publications. Fatality investigation findings led to development of a number of educational materials and journal articles on firefighting hazards such as electrical hazards, and hazards to firefighters involved in live-fire training and working along roadways, to name a few (NIOSH, 2007d, p. 151).

The TI Research Program has written two peer-reviewed publications, and one more has been submitted in the area of improving protection of ambulance workers in patient compartments. There were a number of conference papers and presentations (NIOSH, 2007f, p. A-46). Findings from several fatality investigations of ambulance crashes have been published and are available online. NIOSH also funded several crash investigative reports by NHTSA that are available on the NHTSA Web site.

The TI Research Program produced no peer-reviewed publications in the area of improving protection for emergency workers responding to large-scale disasters and terrorist attacks. The major output of this program was the NIOSH-RAND work on emergency responders. Several NIOSH publications related to disaster safety were also produced.

Outcomes

There are several intermediate outcomes for the TI Research Program's work on emergency responders that demonstrate program impact. End outcomes were not reported for any of the subgoals.

For subgoal 7.1, findings from four FFFIPP investigative reports involving problems with Personal Alert Safety System (PASS) alarms influenced the revision of the NFPA 1982 PASS standard.[45] One FFFIPP investigation was cited in the justification for New York State's Bradley's Law,[46] which prohibits the use of people to play the role of victim under live-fire conditions in fire safety training (NIOSH, 2007d, p. 153). As evidence that the FFFIPP is impacting the workplace, a preliminary evaluation study of the program indicates that recommendations are being used to update safety training programs in thousands of fire departments (NIOSH, 2007d, p. 153). Respirator standards for firefighters and emergency responders developed by NIOSH's National Personal Protective Technology Laboratory researchers were adopted by the Department of Homeland Security in 2004. The standards were also adopted by the NFPA as private-sector consensus standards (NIOSH, 2007d, p. 154).

A major intermediate outcome for subgoal 7.2 has been the contribution of NIOSH project research and TI Research Program anthropometric data to the modification of the Star of Life Ambulance Specification.[47] Among the changes called for was the need for increased head clearance above seats based on NIOSH research. The Star of Life Ambulance Specification is a major driver of ambulance design in the United States. Therefore, this modification is likely to have broad impact. The TI Research Program's work on ambulance safety has also led to continued product development by partners and evaluation of changes to ambulance compartments in testing and concept vehicles. For example, the TI Research Program is working with a Florida fire department to evaluate two ambulances with design changes incorporated in the patient compartments based in part on TI Research Program sled and crash test results. American Medical Response and American Emergency Vehicles developed a concept ambulance that has been used as a demonstrator at EMS conferences that includes restraint systems influenced by

---

[45]The revised standard went into effect on December 20, 2006. In addition to editorial changes, the principal changes to the standard were (1) rewritten design requirements that permit PASS to be other than a "single package," allowing its various components to be located in or combined with other items of protective clothing, and (2) strengthened performance and testing requirements for vibration resistance, impact, water ingress, and alarm signal strength and durability.

[46]Section 1 of the executive law was amended by adding a new section 159-c-1 (Training; live fire conditions) to read:

1. In the training of fire fighters under live fire conditions no person or persons shall play the role of a victim.
2. For purposes of this section, a live fire condition is any unconfined open flame or device that can propagate fire to a building, a training tower, an acquired structure or other combustible material.
3. A violation of this section shall be punishable by a civil penalty not to exceed one thousand dollars paid for by the fire department conducting such training. (NIOSH, 2007f, pp. A44-A45)

[47]The specification revision (KKK 1822 F) was scheduled for full implementation during the third quarter of FY2007.

the TI Research Program and its partners. As an indicator of stakeholder interest, TI Research Program crash test videos have been requested by the International Association of Fire Fighters (IAFF) and other EMS services and fire departments. IAFF incorporated these videos into emergency vehicle operator training for its membership (NIOSH, 2007d, pp. 160-161).

Intermediate outcomes for subgoal 7.3 are in relation to the NIOSH-RAND report on disaster safety management. One intermediate outcome—an American Industrial Hygiene Association 2005 white paper called "Industrial Hygienists' Role and Responsibilities in Emergency Preparedness and Response"[48]—can be tied directly to recommendations from the NIOSH-RAND report. The report's recommendations may have also been an input into information provided in the National Incident Management System (NIMS)[49] (released March 2005) and to a task force updating the National Response Plan. With respect to NIMS, the TI Research Program cites the role of the safety officer[50] (as described in NIMS) as similar to concerns expressed in the NIOSH-RAND report regarding the need to coordinate activities of organizations and agencies that respond to disasters and terrorist incidents (NIOSH, 2007d, p. 168). A reader survey of user response to and the utility of the NIOSH-RAND report found that many responders (emergency directors and administrators for county and municipal agencies) were using the report to inform planning, change plans, modify training, and implement recommendations, among other things (NIOSH, 2007d, p. 168).

Discussion

The TI Research Program's work and transfer activities for emergency responders are responsive to program inputs. All subgoals are in line with NORA I. The committee found the FFFIPP—which draws on NIOSH's strengths of case investigations and surveillance—and the work to develop engineering solutions to problems within ambulance compartments to be particularly strong efforts. Both demonstrate the TI Research Program's ability to work with partners to identify and address problems through interventions.

[48]Available at: http://www.aiha.org/1documents/GovernmentAffairs/EPRWhitePaper_Final.pdf.

[49]NIMS is a system mandated by Homeland Security Presidential Directive 5 that provides a consistent nationwide approach for governments, the private sector, and non-governmental organizations, to work effectively and efficiently together to prepare for, respond to, and recover from domestic incidents, regardless of cause, size, or complexity (DOT, 2008).

[50]According to the NIMS, the safety officer's responsibilities include "ongoing assessment of hazardous environments, the coordination of multiagency safety efforts, and implementation of measures to promote emergency responder safety," and the safety officer "must also ensure the coordination of safety management and functions and issues across jurisdictions, across functional agencies, and with private-sector and nongovernmental organizations" (NIOSH, 2007d, p. 168).

Although the work facilitating the collection of injury and illness surveillance data on emergency responders and the NIOSH-RAND report on emergency responders are admirable, the committee found subgoal 7.3 to be less coherent. This area may be too large for the TI Research Program, particularly when other agencies, such as the Department of Homeland Security, are working on the problem and are likely better resourced for it.

There were no end outcomes for the TI Research Program's work on emergency responders. However, there are a few strong intermediate outcomes, such as the influence of FFFIPP investigation reports on revision of the NFPA 1982 PASS standard and Bradley's Law. The FFFIPP evaluation study finding that recommendations from firefighter fatality investigations are being used to update safety training programs in thousands of fire departments is important, especially considering the risk-accepting culture of firefighters. With respect to ambulance safety, the influence of NIOSH and TI Research Program research on the Star of Life Ambulance Specifications was an important intermediate outcome. Engineering work in this area also appears to have led to much continued research and development by partners.

GOAL 8: REDUCE INJURIES AND FATALITIES TO WORKING YOUTH

The TI Research Program developed three subgoals for research aimed at reducing injuries and fatalities to working youth:

8.1. Influence legislative changes to protect young workers
8.2. Reduce child agricultural injuries
8.3. Foster the development and widespread use of safety materials and intervention strategies to protect young workers[51]

Inputs

Young workers have received special attention from NIOSH for several important reasons. First, young workers appear to have a higher nonfatal injury rate as compared to all workers, when adjusted for the limited hours they work. This is despite the fact that youth are a protected group and restricted from working in the most dangerous jobs (e.g., driving, construction) and doing the most hazardous tasks (e.g., operating heavy equipment). Second, youth have special risk factors, including inexperience on the job, incomplete physical and psychosocial development and, frequently, inadequate training and supervision (especially in

[51]Subgoal 8.3 focused on youth workers in nonagricultural settings.

the informal work sector where many youth first enter the workforce). In addition, society recognizes a special duty to protect our youth.

Laws that place limits on the types of work youth can do have gone unchanged for decades despite changes in the workplace and what is known about occupational safety and health. In 1998, the IOM report *Protecting Youth at Work* called upon the Department of Labor to update these laws based on reviews by NIOSH. The focus of subgoal 8.1 is to increase awareness of the need to update these laws (IOM, 1998; NIOSH, 2007k, p. 173). The work was supported by funding from the DOL Employment Standards Administration (ESA) through an interagency agreement with the TI Research Program to review the adequacy of existing child labor laws that prohibit youth from work identified as especially hazardous (NIOSH, 2007k, p. 174). Subgoal 8.2 addresses the unique problem of injuries among youth working in agriculture. Between 1992 and 2002, rates of fatalities for youth working in agricultural production were 3.5 times greater than rates for youth working in other industries (NIOSH, 2007k, p. 179). Risk of injury among agricultural workers is higher than for other industries, in general, due to the hazardous nature of the work. In 1996, the National Committee for Childhood Agricultural Injury Prevention (of which TI Research Program researchers were members) developed the National Action Plan, which called for NIOSH to serve as the lead federal agency in the prevention of childhood agricultural injury. The initiative was supported by a $5 million congressional appropriation in FY1997. The TI Research Program's implementation plan for the childhood agricultural injury initiative was based on action steps described in the National Action Plan as well as input from stakeholders and peer reviewers (NIOSH, 2007k, p. 180). Subgoal 8.3 was undertaken to increase awareness of safety and health among working youth by developing and implementing effective educational interventions to provide the scientific basis for doing so (NIOSH, 2007k, p. 187). The focus was on reaching all youth, regardless of sector, in order to reduce injuries to youth in their early work lives, and also to invest in the workforce of the future in terms of skills in identifying hazards and preventing injury on the job at any age.

In response to comments made in the 1998 NRC report *Protecting Youth at Work* on the value of fatality investigations for providing the contextual information missing from surveillance systems, the TI Research Program added youth under age 18 as a target for fatality investigations in 1999.

The total intramural and extramural combined funding for the working youth program between FY1997 and FY2005 was $36,013,400 ($14,153,726 intramural and $21,859,674 extramural), making it the largest of the eight goals in terms of funding. At its lowest point in FY1997 the combined funding totaled $1,232,417, and at its peak in FY2001 the funding reached $6,136,918. Funding averaged around $4,000,000 per year. FTEs ranged from a low of 7.7 in FY2005 to a high of 13.3 in FY1999. Most of the funding appears to come from the congressional appropriation of $5,000,000 for child agricultural injury reduction (see Table 1-2). Although the funds do not seem adequate at first glance, it does appear that the available funds have been used effectively, particularly in forming state and local cooperative agreements.

Activities

Since FACE Program's targeting of young worker deaths in 1999, the TI Research Program has conducted 29 FACE investigations of such worker deaths. An additional 42 fatality investigations have been conducted by states with cooperative agreements with the TI Research Program. The committee supports the addition of young worker deaths as a target for the FACE Program, because these investigations provide insight into the circumstances leading to injuries and the necessary intervention strategies. FACE reports may also provide specific guidance in preventing particular types of fatal injuries. In addition, this activity was relevant to subgoal 8.1, because the investigations often found that youth have been injured while performing tasks not prohibited by child labor laws (NIOSH, 2007k, p. 174). In the evidence package, the TI Research Program notes that it is considering scaling back or removing young worker deaths as a FACE Program target, because the investigations often do not identify new strategies for the prevention of young worker injuries (NIOSH, 2007k, p. 176). While scaling back may be warranted, the committee supports continuation of the young worker fatality investigations to respond to changing risk factors.

For subgoal 8.1, the TI Research Program has devoted substantial effort to drawing together existing research results so that they are useful to agencies or legislators considering the development of new rules or laws affecting the safety of young workers. For example, in the interagency agreement with ESA, the TI Research Program performed a review of existing child labor laws that prohibit youth from performing work identified as especially hazardous.[52] The agreement also led to the development of requests for application (RFAs) for extramural research on youth working in construction. A separate RFA for research on childhood agricultural injury resulted in 24 grants for study of the incidence of and risk for childhood agricultural injury (NIOSH, 2007k, p. 174).

While recognizing the importance of child labor laws in protecting young workers, the committee had some reservations about the TI Research Program's emphasis on laws that call for the removal of youth from high-risk jobs rather than

[52]See Miller, M. E., and D. Bush. 2004. Review of the federal child labor regulations: Updating hazardous and prohibited occupations. *American Journal of Industrial Medicine* 45:218-221.

the control of hazards. Although effectively enforced child labor laws can protect working youth, adult workers in these same jobs may continue to be at risk. The committee underscores the continuing need for research to identify and control hazards in jobs commonly held by youth. Many of these same jobs are often held by immigrant and minority workers, who may also be considered vulnerable. Efforts to control hazards driven by concern for youth can result in the protection of all workers.

For subgoal 8.2, it should be noted that most activities under the working youth goal are in the area of agricultural injuries. This is not surprising considering Congress has appropriated several million dollars annually for this research since FY1997. This is an important area, because youth injury rates in agriculture are very high. Despite the small proportion of youth working in agriculture, the agriculture, forestry, and fishing sector accounted for more than 40 percent of all young worker deaths between 1998 and 2002 (NIOSH, 2007k, p. 179). Because of limitations in the BLS Annual Survey of Injuries and Illnesses, the committee cannot make a reliable parallel comparison for nonfatal injuries.

Approximately 25 percent of the congressional appropriation for agricultural injury has been used for surveillance and coordination efforts, and about 75 percent has been used for extramural research and outreach (NIOSH, 2007k, p. 180). Several new surveillance activities were designed to provide an ongoing picture of the incidence of these injuries. Using ongoing survey relationships of NASS, NIOSH has collected information from farm operators about agricultural injuries to children in 1998, 2001, and 2004 (in the CAIS). A parallel survey of minority farms, the M-CAIS, was conducted in 2000 and 2003. In addition, the TI Research Program developed a farm worker injury model to supplement data already collected from farm workers by the DOL in the NAWS. This survey includes young workers, so it provides information about their injuries. NIOSH has also supported several extramural surveillance activities in this area. These include a project that ascertained the number of agricultural injuries incurred by ninth to twelfth graders seen in Minnesota emergency departments. The TI Research Program developed RFAs that resulted in cooperative agreements to fund the National Children's Center for Rural and Agricultural Health and Safety (NCCRAHS) and in 50 research grants (since 1997) to characterize young worker agricultural injuries, risk factors, and outcomes for young agricultural workers and to develop and evaluate interventions.

Since the initiation of the youth FACE investigation program in 1999, the TI Research Program has conducted 3 investigations of young worker deaths in agriculture while the state-based program has conducted 17 such investigations. Findings helped to identify risk factors for fatalities, such as inadequate training and supervision, mismatch between equipment and physical characteristics of youth, use of equipment without safety features, and work assignments prohibited by child labor laws (NIOSH, 2007k, p. 181). As discussed for subgoal 8.1, the TI Research Program led a review of the adequacy of existing child labor laws, some of which are applicable to agricultural work.

Activities for subgoal 8.3 centered on the creation of model approaches and educational materials to provide health and safety training to youth, often through high schools. The TI Research Program developed RFAs for community-based demonstration projects to develop and test approaches and materials that resulted in community-based projects in several regions and states. Through these demonstration projects, many educational materials and dissemination methods were tested and promising approaches were identified and documented, including the development of state teams that bring government agencies and other stakeholders together to coordinate efforts to protect young workers. The TI Research Program's work in this area also involved the establishment of relationships with partners in the public and private sectors for input on the program's products and collaboration in outreach efforts.

Transfer Activities

Youth fatality investigations of both agricultural and nonagricultural injuries conducted by the TI Research Program are posted on the NIOSH Web site and were provided to ESA and OSHA. State-based FACE programs disseminate reports within their states (NIOSH, 2007k, p. 175).

The TI Research Program has devoted considerable time and effort to transfer activities related to subgoal 8.1. The program developed guidelines for the conduct of investigations of young worker deaths and provided technical assistance to states participating in state-based FACE programs. The TI Research Program played an important role in providing statistics and analyses to ESA, which is responsible for developing child labor regulations and developing and enforcing federal child labor laws. In particular, the TI Research Program has provided recommendations for revisions in federal child labor laws based on analyses of existing data and research, some of which were funded by NIOSH. New child labor regulations, based in part on NIOSH input, became effective in 2005 (see "Outcomes," below). The TI Research Program provided technical assistance to state-based young worker injury and illness surveillance systems in Massachusetts and Wisconsin. The TI Research Program also cosponsored the study on the health and safety implications of child labor that led to the IOM (1998) report *Protecting Youth at Work* (NIOSH, 2007k, pp. 173-175).

For subgoal 8.2, the TI Research Program has engaged in transfer activities to facilitate information sharing relevant to reducing injuries of young workers in agriculture. The TI Research Program helped to plan and participated in a 2001 Summit on Childhood Agricultural Injury Prevention that assessed the progress and update of the 1996 National Action Plan. Between 1999 and 2004, the TI Research Program convened an annual meeting of researchers with NIOSH funding through NCCRAHS to share findings and discuss research problems and solutions. The TI Research Program also formed and chairs the federal IAWG on Preventing Childhood Agricultural Injuries. The working group is a forum for those with a stake in childhood agricultural injury prevention to discuss the state of the research and the incorporation of findings into safety products (NIOSH, 2007k, p. 181). The TI Research Program has also been engaged with many organizations—both government and private—to disseminate information about risks to young agricultural workers and how to prevent these risks. The TI Research Program worked with NASS to provide farm safety materials focusing on youth to the approximately 100,000 farm operators that it surveys (NIOSH, 2007k, p. 182). Finally, as stated for subgoal 8.1, the TI Research Program provided recommendations to ESA for revisions to federal child labor laws by submitting comments to an Advanced Notice for Proposed Rulemaking. These comments included a recommendation for increasing the minimum age for hazardous agricultural work from 16 to 18 and the removal of the family farm exemption. The 2002 NIOSH report recommended revisions to 8 or 11 HOs in agriculture and a new HO for both agricultural and nonagricultural industries prohibiting youth from work that requires respirators (NIOSH, 2007k, p. 183).

Transfer activities for subgoal 8.3 included a provision of technical assistance to the community-based demonstration project grantees in the execution of projects and a modification of materials developed by these programs for national use (NIOSH, 2007k, p. 187). Thousands of copies of NIOSH publications on young worker safety were disseminated through targeted mailings.

Outputs

The working youth program has produced a number of tangible, relevant, and helpful outputs. For example, for subgoal 8.1, NIOSH not only used existing surveillance data but also developed new surveillance sources, both by building on existing surveys conducted by other agencies and by funding new surveillance activities. These include new state-based surveillance activities in Wisconsin and Massachusetts focused on working youth. They also include sector-based surveillance in agriculture and construction. This expanded surveillance is paving the way for a more comprehensive understanding of the hazards facing working youth. These outputs will enable NIOSH to better plan new research focused on working youth and to better evaluate the impact of ongoing activities.

Specifically for subgoal 8.1, the TI Research Program has written 15 peer-reviewed articles since 1996, which have largely been concerned with establishing the magnitude and patterns of young worker injuries and fatalities. External collaborators have generated 25 articles in addition to this. The TI Research Program authored 11 NIOSH documents with data on young worker injuries and completed 29 FACE reports of investigations of youth fatalities, while states with cooperative agreements completed another 42 investigation reports. Findings from the investigations helped to show that work tasks not prohibited by existing child labor laws have resulted in fatal injuries. NIOSH has also been actively working with collaborators to produce recommendations concerning 34 HOs. Finally, the TI Research Program's contribution to young worker initiatives in Massachusetts and California has led to a broadening effect involving collaborations with multiple other agencies. This appears to be a highly successful model of transfer that has been replicated now by multiple states with the support of the National Young Workers Resource Center (operated jointly by the California and Massachusetts programs and funded by OSHA).

For subgoal 8.2, the TI Research Program has published 23 articles in the peer-reviewed literature as well as 31 articles by external researchers. These articles describe the magnitude and patterns of injuries among young agricultural workers and make recommendations for prevention. In addition, nine NIOSH and three USDA documents have been generated (NIOSH, 2007k, p. 183). There have also been 3 fatality investigation reports completed by the TI Research Program involving young worker deaths in agriculture, as well as 17 fatality investigation reports completed by states. In 1997, the TI Research Program formed the federal IAWG on Preventing Childhood Agricultural Injuries. The TI Research Program chairs the working group, which meets biannually, and provides a forum for discussion of ongoing activities and the incorporation of new findings into safety products (NIOSH, 2007k, p. 181). The TI Research Program provided recommendations to ESA for revisions to federal child labor laws by submitting comments to an Advanced Notice for Proposed Rulemaking and in the 2002 NIOSH report, several of which applied to youth working in agriculture.

The TI Research Program has published 10 peer-reviewed articles since 1996 supplemented by another 23 articles from external researchers for work related to subgoal 8.3. Thirteen NIOSH documents have also been generated. Material has also been generated to engage young workers directly, including stand-alone curricula and curricular activities for integration into current high school programs. NIOSH, in collaboration with the University of California at Berkeley program, has now developed curricula tailored for each of the 50 states and posted them on its Web site, making it easy for educators to access and use these educational approaches.

Outcomes

An intermediate outcome cited for subgoal 8.1 was changes to FLSA child labor laws that went into effect February 14, 2005.[53] Because research and recommendations from the TI Research Program were cited among justifications for the changes in the final rule,[54] the committee agrees that the TI Research Program's work had some role in shaping the changes to the child labor laws. Another intermediate outcome was broad stakeholder use of the 2002 report "National Institute for Occupation Safety and Health (NIOSH) Recommendations to the U.S. Department of Labor for Changes to Hazardous Orders,"[55] which echoed the comments submitted by NIOSH to the DOL in the rule-making comment period. In addition to the DOL, the report has been used by others looking to improve workplace safety for youth. In 2003, a California congressman used the report as the basis for a section of a proposed Youth Worker Protection Act (H.R. 2870, 109th Congress) on hazardous occupations for youth, although the legislation was not passed. The Child Labor Coalition referenced the report in a 2006 letter to the Secretary of Labor requesting action on child agricultural labor regulations and in a document questioning U.S. compliance with International Labor Convention 182, which calls for prohibition and elimination of certain forms of child labor. Another intermediate outcome for subgoal 8.1 was the contribution of a youth FACE investigation conducted in Oregon to a 2005 child labor law in that state prohibiting youth under age 18 from working in occupations involving the use of explosives (NIOSH, 2007k, pp. 175-176). A NIOSH-funded project in Massachusetts—the Teens at Work Injury Surveillance and Prevention Project—was instrumental in providing data that contributed to a revision of the Massachusetts child labor laws as well as changes in policy and training in vocational education schools.

The TI Research Program cites the use of surveillance data on youth working in agriculture performed by other parties both in government and in the private sector to guide prevention efforts as an intermediate outcome for subgoal 8.2. A particularly strong indication of stakeholder interest was the use of TI Research

[53]Changes were for youth workers and (1) cooking and cooking-related duties, (2) all work on or about roofs, (3) loading of power-driven balers and compactors, and (4) teen driving on the job. A complete description of changes can be found in the final rule at http://www.dol.gov/esa/regs/fedreg/final/2004027182.htm.

[54]The final rule states that NIOSH provided the DOL with epidemiological data on a number of issues related to regulations for youth workers aged 14 and 15 and occupations covered by HOs during the rule-making comment period. NIOSH also provided the DOL with statistics regarding occupational injuries and made several recommendations (*Federal Register* 69(241):75381 (2004)).

[55]Document can be found at: http://www.cdc.gov/niosh/docs/NIOSHRecsDOLHaz/pdfs/DOL-recomm.pdf.

Program statistics on injuries among youth working on farms in the proposed Children's Act for Responsible Employment (CARE) of 2005[56] (CARE Act, H.R. 3482). This act proposed changes to child labor laws in agriculture and identified the TI Research Program youth farm injury data as a source of data for an annual report on injuries to youth working on farms in the United States. The 2002 NIOSH report that provided recommendations to ESA on changes to child labor laws has been used by others advocating for changes in child labor regulations in the agriculture industry (NIOSH, 2007k, p. 183).

The TI Research Program also notes an end outcome for subgoal 8.2—namely, a decline in the number of injuries and the rates of nonfatal injuries among agricultural workers under age 18 between 1998 and 2004. The number of work-related injuries decreased 50 percent over this period (from 11,970 in 1998 to 5,740 in 2004), and the rates of nonfatal, work-related injuries decreased 30 percent (from 10.5 per 1,000 working youths in 1998 to 7.3 per 1,000 working youths in 2004; NIOSH, 2007k, p. 184).[57] The committee believes that the contributions of the TI Research Program, in particular through outreach activities done in collaboration with partners, likely played some role.

As an intermediate outcome for subgoal 8.3, young worker safety efforts have continued for several state-based teams. Work of investigators with community-based demonstration projects in Oakland, California, and the Northeast Young Worker Resource Center has been extended through funding from OSHA (NIOSH, 2007k, p. 189). The Teens at Work surveillance project in Massachusetts resulted in a technological intervention. The project identified hot coffee and slurry from coffee brew baskets as a common source of occupational injury among Massachusetts teens. In response to the finding, the corporate headquarters of a national bakery chain in 2001 began to require that store owners purchasing new equipment install brew baskets and shields to prevent spillage. A retrofit kit was designed and made available to store owners who are not purchasing new equipment (NIOSH, 2007k, p. 189). Stakeholder interest in the TI Research Program's work in this area has been evidenced by requests for additional copies of TI Research Program-developed young worker safety and health materials; incorporation of materials into school-based occupational safety and health classes; and citation of TI Research Program statistics, findings, and prevention recommendations by the press and in safety programs (NIOSH, 2007k, p. 190).

---

[56]The act was submitted in the House of Representatives in July 2005, but never passed.

[57]The discrepancy between the rate and the absolute number derives from decreasing numbers of youth working in agriculture.

Discussion

The TI Research Program's activities for young workers have demonstrated the significance of innovative, as well as continuing, surveillance activities. The value of such surveillance data to focused approaches cannot be overemphasized in terms of the benefits likely to accrue. In other areas of research activity, however, it is clear that many activities are better described as research to practice (r2p). While the committee endorses the TI Research Program's efforts to work with various constituencies to implement products and programs, it appears that there is a significant danger that actual research into the causes of traumatic injury is underemphasized relative to implementation. The choice to add youth to the FACE Program in 1999 was a good one, and the committee supports continuation of a scaled-down youth FACE Program in the future if new interventions are no longer being identified.

The young worker activity also focused on the development of legislative approaches to reducing injuries. The moderate nature of its success apparently arises from a significant lack of interest and "pickup" by the various bodies responsible for promulgation and, perhaps more importantly, enforcement. The ancillary activity of partnering with groups representing the constituency at risk, however, appears to have been much more successful and demonstrated a "pull" for these activities, be they r2p or research, rather than the "push" that NIOSH is (appropriately) trying to avoid.

Activities resulted in a number of relevant outputs. Particularly strong was the development of new surveillance sources that add to the knowledge base of risk factors for occupational injury among youth. More comprehensive data will help NIOSH and the TI Research Program in the evaluation of existing efforts as well as the targeting of future efforts toward youth. A number of products developed and tested through the demonstration projects seem to have been well received by stakeholders, as evidenced by requests for additional copies and incorporation of products into youth occupational safety programs.

Finally, the TI Research Program's goal on working youth appears to have resulted in several modest yet important intermediate outcomes. Outputs, such as surveillance data and the 2002 NIOSH report "National Institute for Occupation Safety and Health (NIOSH) Recommendations to the U.S. Department of Labor for Changes to Hazardous Orders"[58] summarizing recommendations for updating child labor laws, have continued to be used by groups both in the government and in the private sector. Through outreach activities with partners, the TI Research Program in all probability had some part in the end outcome for subgoal 8.2,

[58]Document can be found at: http://www.cdc.gov/niosh/docs/NIOSHRecsDOLHaz/pdfs/DOL-recomm.pdf.

namely a decline in injuries among young agricultural workers between 1998 and 2004.

EVALUATION OF RELEVANCE

The guidance for assessing relevance in the Framework Document is "whether the program appropriately sets priorities among research needs" and "how engaged the program is in appropriate transfer activities" (see Appendix A, p. 137). A relevance score of 5 signifies a conclusion that the research is in high-priority areas and the program is significantly engaged in transfer activities, whereas a score of 4 signifies that the research is in priority subject areas and the program is engaged in appropriate transfer activities (see Box 2-1).

The committee reviewed the work supporting the 8 specific goals (including the 19 subgoals) that constitute the TI Research Program. The goals represent a mix of long-standing safety concerns (e.g., agricultural injuries), newer or emerging areas of emphasis for the TI Research Program (e.g., falls from telecommunications towers), and attention driven by congressional directive (e.g., the AFS). Three goals represent specific worker populations identified by location (Alaska), age (youth), or sector (emergency response). NIOSH has clearly driven a national sensitivity to some specific safety problems. For example, the committee concludes that NIOSH attention to workplace violence has shined a light on a previously neglected area.

The committee concludes that for the most part, the goals are appropriate and relevant to the burden of traumatic injury in the workplace. The burden of injury

BOX 2-1
Scoring Criteria for Relevance

5 = Research is in high-priority subject areas and NIOSH is significantly engaged in appropriate transfer activities for completed research projects or reported research results.
4 = Research is in priority subject areas and NIOSH is engaged in appropriate transfer activities for completed research projects or reported research results.
3 = Research is in high-priority or priority subject areas, but NIOSH is not engaged in appropriate transfer activities; or research focuses on lesser priorities but NIOSH is engaged in appropriate transfer activities.
2 = Research program is focused on lesser priorities and NIOSH is not engaged in or planning some appropriate transfer activities.
1 = Research program is not focused on priorities and NIOSH is not engaged in transfer activities.

represented by the eight major goal areas is certainly high, although the committee did not attempt to independently assess the burden of injuries in all occupations or work sites in the country as part of its review. Rather, the committee understands the challenges NIOSH faces in prioritizing research with such restricted resources and concludes that given this limited budget and based on committee members' expert judgment, the TI Research Program has made overall appropriate selections of general areas to pursue.

While the committee concluded that many of the goal areas were high priority—such as Alaska and falls from elevations—it identified gaps, such as falls from same elevation, several sectors within workplace violence, and a narrow focus within machines. There were also areas in which TI Research Program research was "stuck" in its approach (e.g., a focus on engineering solutions to tractor rollovers, instead of a switch to policy research to accomplish transfer to the field).

As described in previous sections of this chapter, the TI Research Program engages in appropriate transfer activity within some, but not all, of the goal areas. TI Research Program staff has participated in the ANSI Z15 committee (subgoal 1.1), an important collaborative transfer activity. The transfer activities in goal 2, if successful, are likely to lead to the introduction of TI Research Program-dependent safety innovations, primarily engineering-based, but also developed from FACE investigations. TI Research Program staff has urged ASABE to develop a performance standard for the AutoROPS (subgoal 4.1), it has contributed to a DOL-OSHA committee that resulted in OSHA guidelines for nursing homes (subgoal 5.1), it has engaged in significant transfer activities related to legislative changes to protect young workers (subgoal 8.1), and it has provided recommendations for revisions in federal child labor laws (which became effective in 2005). The TI Research Program-funded Youth at Work program was instrumental in the passage of revised child labor laws in Massachusetts. These transfer activities are appropriate and related to the intermediate outcomes demonstrated for some of the program activities.

Score for Relevance

In summary, the committee notes impressive work, including transfer, in priority goal areas. **The committee assigns a score of 4 for the relevance of the TI Research Program.** This score is based on the fact that research is mostly in priority and some high-priority subject areas and has a range of involvement in transfer activities.

BOX 2-2
Scoring Criteria for Impact

5 = Research program has made major contribution(s) to worker health and safety on the basis of end outcomes or well-accepted intermediate outcomes.
4 = Research program has made some contributions to end outcomes or well-accepted intermediate outcomes.
3 = Research program activities are ongoing and outputs are produced that are likely to result in improvements in worker health and safety (with explanation of why not rated higher). Well-accepted outcomes have not been recorded.
2 = Research program activities are ongoing and outputs are produced that may result in new knowledge or technology, but only limited application is expected. Well-accepted outcomes have not been recorded.
1 = Research activities and outputs do not result in or are NOT likely to have any application.
NA = Impact cannot be assessed; program not mature enough.

EVALUATION OF IMPACT

The directions for assessing impact on end outcomes or well-accepted intermediate outcomes in the framework seem to hinge on a distinction between "major contributions" for a score of 5 and "some contributions" for a score of 4 (see Box 2-2). The committee understands that a determination of major contributions does not "imply that the NIOSH program was solely responsible for observed improvements in worker health and safety"; the committee is aware of the challenges in assigning such responsibility and has not attempted to do so.

Intermediate Outcomes

The Framework Document describes intermediate outcomes as "important indicators of stakeholder response to NIOSH outputs. They reflect the impact of program activities and may lead to the desired end outcome of improved workplace safety and health." Intermediate outcomes fall into the major categories of public policy impact, training and education, and self-reported use or repackaging of NIOSH material by stakeholders. The committee found evidence of effects on well-established intermediate outcomes in all of the eight goal areas and in all types of outcomes. These have been documented for each goal and subgoal in previous sections of this chapter. This section briefly summarizes some exemplars by the type of intermediate outcome rather than by goal.

Intermediate outcomes related to **public policy impact** refer to the production (by those outside NIOSH) of guidelines or regulations based wholly or partly on NIOSH research. The TI Research Program has been associated with the development of both voluntary guidelines and regulations in several areas (goal 1: motor vehicles; goal 2: falls from elevations; goal 3: workplace violence; goal 4: machines; goal 5: back injuries; goal 6: Alaska; and goal 8: working youth).

Approval of the ANSI Z15.1 standard, Safe Practices for Motor Vehicle Operations, is a success for goal 1. This filled the gap left by the Federal Motor Carrier Safety Regulations, which apply only to commercial drivers. The ANSI standard will help protect all employees who operate a motor vehicle as part of their job. Another very important intermediate outcome was the incorporation of a NIOSH recommendation related to HO No. 2 (motor vehicle occupations) into a final rule published by the DOL in 2004. The DOL published a final rule[59] on the HO prohibiting all workplace driving on public roadways by 16-year-old workers and placed restrictions on driving by 17-year-old workers.

The TI Research Program's work on falls from telecommunications towers supported the development of two important policy documents—namely, the first statewide telecommunications tower standard (in North Carolina), which addresses safety procedures to be used during tower construction and maintenance, and the OSHA Telecommunications Tower Task Force Compliance Directives, which address inspection procedures. OSHA published recommendations for workplace violence prevention programs in 1998 and guidance documents for preventing workplace violence to healthcare and community service workers in 2004. Notable intermediate outcomes in goal 4 (machines) are the revisions to ANSI Z245.5 Standards Committee to require a security switch. This will serve to protect workers at municipal and commercial recycling centers. An amendment to the FLSA restricts workers under the age of 18 years from loading balers that do not meet the ANSI standard and prohibits them from operating balers (HO No. 12).[60]

The research in patient handling has been cited as an important source of evidence leading to the passage of safe patient handling legislation in several states (Texas, Washington, Rhode Island, Ohio, Hawaii, and New York). As discussed above, BLS data indicate a decrease in rates of injuries likely related to patient handling from the early 1990s to the present. Other changes in public policy related to TI Research Program findings are the USCG Dockside Pre-season Boarding

[59]Regulations implementing the act permit the Secretary of Labor to prohibit the employment of youth in occupations declared "particularly hazardous for the employment of children . . . or detrimental to their health or well-being" (29 USC 201 Sec. 3(1)). These prohibited activities are referred to as Hazardous Occupations Orders (HOs). The minimum age, by statute, for HOs in nonagricultural occupations is 18; the minimum age in agricultural occupations is 16.

[60]HO No. 12 (balers), 29 CFR Part 570.63.

Program and the helicopter safety guidelines related to goal 6, and changes in the FLSA in 2005 related to goal 8.

Intermediate outcomes related to **training and education** refer to programs sponsored by other organizations. These have been documented in previous sections of this chapter and are summarized only briefly here. The NIOSH document "Building Safer Highways and Work Zones"[61] has been used in an OSHA training course and by other organizations for safety training. TI Research Program work on falls from telecommunications towers supported the development of OSHA train-the-trainer programs as well as comprehensive NATE manuals. TI Research Program staff and their research related to workplace violence have influenced OSHA recommendations and guidelines as well as guides on workplace violence prevention by the Office of Personnel Management and the American Society for Industrial Security. TI Research Program findings on safe patient handling have been used in professional organization training programs and textbooks. Work in Alaska aviation has been used extensively in FAA training programs, including public service announcements and training videos

Another category of intermediate outcome is **self-reported use and/or repackaging** by stakeholders. Many NIOSH reports have been picked up by stakeholders and disseminated. For example, the TI Research Program-authored report "Building Safer Highway Work Zones: Measures to Prevent Worker Injuries from Vehicles and Equipment"[62] has been distributed by OSHA offices and reproduced as part of the Laborers' Health and Safety Fund of North America 2003. The blind area equipment reports have been requested by several stakeholders.

The committee includes in this category stakeholder-initiated changes in safety technology and workplace environments.[63] The FACE Program can be particularly useful in illuminating these kinds of risks. A FACE investigation of a scalping accident and the subsequent identification of other cases led to a retrofit by the manufacturer of the hay baler involved in all five cases. The manufacture of sealed agricultural plow frames revised its manufacturing process to avoid the scrap metal that appeared to be responsible for ignition hazards identified in FACE investigations. Although the original ROPS work occurred outside the period of committee evaluation, manufacturers have changed the design of tractors (although uptake of the technology is quite slow).

Equally important to the research that led to improvements in equipment design is research showing that a technology is not effective. The research on back

[61]NIOSH. 2001. Building safer highway work zones: Measures to prevent worker injuries from vehicles and equipment. HHS (NIOSH) Publication No. 2001-128.

[62]See footnote 61 for reference.

[63]This is distinct from public policy impact, which would force stakeholders to make changes.

belt use for materials handling demonstrated no effect on injuries, and two major retail chains no longer provide or require back belt use. Research demonstrating no effect of an intervention is as important as research that does show an effect, because discontinuance of an ineffective intervention allows more promising interventions to be studied and introduced, which might not have occurred had the workers and their management assumed the technology to be effective.

Other safety technology introductions based on the research of the TI Research Program, although limited, include those involving patient compartments, such as the Star of Life Ambulance Specifications, changes in the patient compartment in some ambulances in Florida, and restraint systems for a concept ambulance.

End Outcomes

NIOSH reports improvements in end outcomes for five of the eight goals: workplace violence (a very modest decrease in fatalities compared to societal homicides), machines, acute back injury in patient handling, high-risk occupations in Alaska, and nonfatal agricultural work-related injuries in youth (but not non-agricultural work-related injuries in youth).

As noted in this list, the committee greets with some reservation the reports of an impact on workplace violence fatalities given that societal homicides decreased similarly over the same period of time and these trends likely are inextricably linked. The committee discusses important contributions to intermediate outcomes in a previous section.

NIOSH reports a steady decrease in fatalities caused by machines, plant and industrial powered vehicles, and tractors since 1992. The number of deaths decreased 16 percent and the fatality rates declined 30 percent. The committee agrees with the NIOSH conclusion that it is difficult to quantify the contributions of the TI Research Program to these decreases. The probable source of improvement in end outcomes related to goal 4 (machines) is primarily due to the FACE investigations (subgoal 4.3), not the other recent efforts on ROPS or the limited work on paper balers. The committee is supportive of the FACE Program and recognizes its contributions. It raises this merely to make the point that the end outcomes related to the machines program derive from one program and not from other interventions supported in subgoals 4.1 and 4.2.

There is little doubt that the successes in the Alaska program are numerous and important. The program benefited from a dedicated source of funding (a congressional directive), a reasonably narrow focus, and locus as a state-based program. Having TI Research Program staff in close physical proximity to the many important stakeholders was likely influential in the success of the program. This is not a feasible strategy in most areas of TI Research Program work, however. As an aside, the committee hopes that the TI Research Program can extend the AFS successes in Alaska and across the country. The combination of important partnerships, focus, sufficient resources, and a state-based program with specific attention to the public health model could be applied to other high-risk industries in Alaska as well as similar high-risk industries in other parts of the country. The AFS model can serve as a model of success for the program as a whole.

Surveys from farm operators suggest a marked decrease in nonfatal work-related injuries among youth between 1998 and 2004. There is not, however, evidence of decreased injuries of working youth in other sectors.

The committee views the changes in end outcomes with some reservations, unsure of the TI Research Program contributions to the results. However, the committee favorably views any progress in reducing occupational traumatic injuries and is confident of some degree of NIOSH contributions. Given the very limited resources under which the TI Research Program operates, results regarding end outcomes are admirable. Other important external factors that likely restrain success include OSHA inaction, the difficult nature of some of the industries in which injuries are high (e.g., the culture of individualism and risk acceptance in the agricultural and logging industries), and the difficulties of getting small- to medium-sized enterprises to invest in safety technologies. The TI Research Program might benefit from efforts to increase its capacity and expertise in policy research, in behavioral and social science research, and in directed efforts to understand how to impact small- to medium-sized enterprises. Chapters 3 and 4 address these issues.

Score for Impact

The committee commends the TI Research Program for its contributions toward reducing occupational traumatic injuries. As documented in the previous section, the TI Research Program is associated with impact on either intermediate or end outcomes in each major goal. The committee recognizes that external factors, specifically severely limited resources and inaction on the part of OSHA, can be significant barriers to the TI Research Program's progress in some goal areas. However, the committee notes (1) the lack of demonstrated effect on end outcome data in three goal areas and in some subgoals in the other five goals, (2) the inability to determine what degree of responsibility the TI Research Program bears for the documented improvements in end outcomes or for the intermediate outcomes, and (3) a lack of significant intermediate outcomes for some subgoals. **The committee assigns a score of 4 for the impact of the TI Research Program.**

REFERENCES

ASIS (American Society of Industrial Security). 2006. Guidelines for WPV. http://www.asisonline.org/guidelines/guidelineswpvfial.pdf (accessed July 23, 2008).

Azaroff, L. S., C. Levenstein, and D. H. Wegman. 2002. Occupational injury and illness surveillance: Conceptual filters explain underreporting. *American Journal of Public Health* 92(9):1421-1429.

BLS (Bureau of Labor Statistics). 2008. Injuries, illnesses, and fatalities. http://www.bls.gov/iif/home.htm (accessed July 23, 2008).

Bobick, T. G., J. L. Belard, H. Hsiao, and J. T. Wassell. 2001. Physiological effects of back belt wearing during asymmetric lifting. *Applied Ergonomics* 32(2001):541-547.

Conway, H., and J. Svenson. 1998. Occupational injury and illness rates, 1992-1996: Why they fell. *Monthly Labor Review* November:36-58.

DOL (Department of Labor). 2008. OSH Act of 1970 Sec. 5 duties. http://www.osha.gov/pls/oshaweb/owadisp.show_document?p_id=3359&p_table=OSHACT (accessed July 31, 2008).

DOT (Department of Transportation). 2008. Glossary. http://ops.fhwa.dot.gov/publications/evac_primer/26_glossary.htm (accessed July 24, 2008).

Giorcelli, R. J., R. E. Hughes, J. T. Wassell, and H. Hsiao. 2001. The effect of wearing a back belt on spine kinematics during asymmetric lifting of large and small boxes. *Spine* 26(16):1794-1798.

Haddon, W. 1968. The changing approach to the epidemiology, prevention, and amelioration of trauma: The transition to approaches etiologically rather than descriptively based. *American Journal of Public Health* 58:1431-1438.

Haddon, W. 1972. A logical framework for categorizing highway safety phenomena and activity. *Journal of Trauma* 12(1):193-207.

Hallman, E. M. 2005. ROPS retrofitting: Measuring effectiveness of incentives and uncovering inherent barriers to success. *Journal of Agricultural Safety and Health* 11(1):75-84.

Hendricks, S. A., E. L. Jenkins, and K. R. Anderson. 2007. Trends in workplace homicides in the U.S., 1993-2002: A decade in decline. *American Journal of Industrial Medicine* 50:316-325.

IOM (Institute of Medicine). 1998. *Protecting youth at work: Health, safety, and development of working children and adolescents in the United States.* Washington, DC: National Academy Press.

Jenkins, E. L. 1996. Current Intelligence Bulletin 57, Violence in the workplace: Risk factors and prevention strategies. HHS (NIOSH) Publication No. 96-100.

Leigh, J. P., J. Marcin, and T. R. Miller. 2004. An estimate of the U.S. government's undercount of nonfatal occupational injuries. *Journal of Occupational and Environmental Medicine* 46(1):10-18.

Merchant, J., and J. Lundell. 2001. Workplace violence intervention research workshop. *American Journal of Preventative Medicine* 20(2):135-140.

Myers, J. R., and K. A. Snyder. 1995. Roll-over protective structure use and the cost of retrofitting tractors in the United States, 1993. *Journal of Agricultural Safety and Health* 1(3):185-197.

NIOSH (National Institute of Occupational Safety and Health). 2006. NIOSH awards funding to ag research centers for initiative to reduce tractor deaths, injuries. NIOSH Update. March 21, 2006. http://www.cdc.gov/niosh/updates/upd-03-21-06.html (accessed July 24, 2008).

NIOSH. 2007a (unpublished). *Overview of NIOSH.* In the evidence package provided to the Committee to Review the NIOSH TI Research Program. NIOSH.

NIOSH. 2007b (unpublished). *Overview of the TI Research Program.* NIOSH.

NIOSH. 2007c (unpublished). *Current TI research goals and sub goals: Reduce injuries and fatalities due to motor-vehicles.* In the evidence package provided to the Committee to Review the NIOSH TI Research Program. NIOSH.

NIOSH. 2007d (unpublished). *Current TI research goals and sub goals: Reduce injuries and fatalities to emergency responders.* In the evidence package provided to the Committee to Review the NIOSH TI Research Program. NIOSH.

NIOSH. 2007e (unpublished). *Current TI research goals and sub goals: Reduce injuries and fatalities due to falls from elevations.* In the evidence package provided to the Committee to Review the NIOSH TI Research Program. NIOSH.

NIOSH. 2007f (unpublished). *Appendix 1: Supporting evidence for the TI goals/subgoals.* In the evidence package provided to the Committee to Review the NIOSH TI Research Program. NIOSH.

NIOSH. 2007g (unpublished). *Current TI research goals and sub goals: Reduce injuries and fatalities due to workplace violence.* In the evidence package provided to the Committee to Review the NIOSH TI Research Program. NIOSH.

NIOSH. 2007h (unpublished). *Current TI research goals and sub goals: Reduce injuries and fatalities due to machines.* In the evidence package provided to the Committee to Review the NIOSH TI Research Program. NIOSH.

NIOSH. 2007i (unpublished). *Current TI research goals and sub goals: Reduce acute back injury.* In the evidence package provided to the Committee to Review the NIOSH TI Research Program. NIOSH.

NIOSH. 2007j (unpublished). *Current TI research goals and sub goals: Reduce injuries and fatalities among workers in Alaska.* In the evidence package provided to the Committee to Review the NIOSH TI Research Program. NIOSH.

NIOSH. 2007k (unpublished). Current TI research goals and sub goals: Reduce injuries and fatalities to working youth. In the evidence package provided to the Committee to Review the NIOSH TI Research Program. NIOSH.

NIOSH. 2008. Fatality Assessment and Control Evaluation (FACE) Program. http://www.cdc.gov/niosh/face/ (accessed July 23, 2008).

OPM (Office of Personnel Management). 1998. Dealing with workplace violence: a guide for agency planners. OWR-09, February 1998.

Pana-Cryan, R., and M. L. Myers. 2000. Prevention effectiveness of roll-over protective structures—Part III: Economic analysis. *Journal of Agricultural Safety and Health* 6(1):57-70.

Pana-Cryan, R., and M. L. Myers. 2002. Cost effectiveness of roll-over protective structures. *American Journal of Industrial Medicine* 42(S2):68-71.

Pratt, S. G., D. E. Fosbroke, and S. M. Marsh. 2001. Building safer highway work zones: Measures to prevent worker injuries from vehicles and equipment. Cincinnati, OH: U.S. Department of Health and Human Services, Public Health Service, Centers for Disease Control and Prevention, National Institute for Occupational Safety and Health. HHS (NIOSH) Publication No. 2001-128.

Rosenman, K. D., A. Kalush, M. J. Reilly, J. C. Gardiner, M. Reeves, and Z. Luo. 2006. How much work-related injury and illness is missed by the current national surveillance system? *Journal of Occupational and Environmental Medicine* 48(4):357-365.

Stout, N., and C. Bell. 1991. Effectiveness of source documents for identifying fatal occupational injuries: A synthesis of studies. *American Journal of Public Health* 81(6):725-728.

Wassell, J. T., L. I. Gardner, D. P. Landsittel, J. J. Johnston, and J. M. Johnston. 2000. A prospective study of back belts for prevention of back pain and injury. *Journal of the American Medical Association* 284(21):2727-2732.

3

Targeting of New Research and Identification of Emerging Research Needs

The second part of the committee's charge was to perform an assessment of the Traumatic Injury (TI) Research Program's effectiveness in targeting new research areas and identifying emerging issues most relevant to future improvements in workplace protection. The first part of this chapter outlines the processes that the National Institute for Occupation Safety and Health (NIOSH) and the TI Research Program use for research project planning and resource allocation as well as identification of emerging issues relevant to the NIOSH mission. The second part of this chapter is the committee's evaluation of these processes. The chapter concludes with the committee's evaluation of the TI Research Program's draft "strategic goals for the future." This evaluation includes comments on the relevance of the proposed strategic goals, identifies gaps in the proposed research and emerging issues, and makes other suggestions about how the goals can be strengthened in later iterations.

THE TI RESEARCH PROGRAM'S PROCESS FOR TARGETING NEW RESEARCH AND IDENTIFYING EMERGING RESEARCH NEEDS

The following descriptions of the TI Research Program's process for research project planning and resource allocation and identification of emerging research needs come primarily from the evidence package provided to the committee. The majority of the information relates to intramural research, because the committee did not have available much information on how NIOSH or the TI Research Program develops research priorities for extramural research.

Research Project Planning and Resource Allocation[1]

NIOSH TI research project planning takes place at the research division level and at the institute level.

At the division level, Division of Safety Research (DSR) staff propose research projects within the context of program drivers, which may include surveillance findings on injury incidence and severity, worker groups with the greatest numbers and risk of death or injury, congressional mandates, stakeholder input, or research needs outlined in the 1998 National Occupational Research Agenda (NORA)-TI team white paper (NIOSH, 1998). DSR leadership—with input from staff—rates and ranks new project concepts based on project need, soundness of approach or methods, and expected impact (NIOSH, 2007a, p. 43). Concepts that rank highest are approved based on funding availability. Staff may then develop research protocols within the approved concept areas. Research protocols are peer-reviewed internally, and may also be presented at public meetings for stakeholder input and to assess the interest in and potential impact of the research. When research projects end, funding returns to a pool for competition for new project concepts proposed by DSR staff (NIOSH, 2007a, p. 43).

At the institute level, NIOSH researchers, on an annual basis, have the opportunity to compete for NORA funds set aside for intramural research. Competition for these funds is institute-wide, although the NIOSH director may sometimes call for focus in a specific area. The director makes final project funding decisions using scores from an external peer review and available funding. When projects are completed, funding returns to the NORA pool for renewed intramural competition (NIOSH, 2007a, p. 43).

According to DSR leadership, most current TI Research Program projects are funded through the institute-wide NORA funding competition. DSR "base" funds (annual division or lab allocation) have diminished and are now used primarily for ongoing surveillance and field investigation programs, as well as congressionally mandated projects (NIOSH, 2007a, p. 43). Funds left over from the few DSR-specific funded projects that end each year are used to cover costs to the Centers for Disease Control and Prevention, which have increased in recent years, as well as annual cost-of-living increases. There has not been competition for new projects with DSR base funds for the past 3 years.[2]

Extramural researchers may request NIOSH funding for TI research through NIOSH general program announcements, targeted requests for applications (RFAs) developed by the TI Research Program aimed at filling specific program gaps, and

---

[1]This description does not include project planning and resource allocation for mining TI research, which is managed by the separate NIOSH Mining Safety and Health Research Program.

[2]Personal communication from N. Stout.

cooperative agreements. As with the institute-wide competition for NORA funds, extramural applicants submit proposals that are externally peer-reviewed and scored (NIOSH, 2007a, p. 43).

As part of a process of continued project assessment and adjustment, quarterly progress reports for each intramural TI Research Program project are distributed among staff for information sharing, progress assessment, and input. There is also an annual midyear review of programs and projects. Projects found not to be meeting anticipated progress or value may be discontinued. These assessments have been important in ensuring the sustained quality and relevance of TI Research Program's work (NIOSH, 2007a, pp. 43-44).

NIOSH is currently implementing a matrix organizational process for coordinating research projects across programs and divisions or labs. The goals of this organizational process are to improve internal management and coordination of intramural and extramural research and planning, and to increase research relevance and impact. Project administration and management remains a division-level responsibility (NIOSH, 2007a, p. 44).

Stakeholder Input

The TI Research Program has several means by which it receives input from stakeholders on its research programs. For intramural projects, NIOSH frequently holds a public meeting announced in the *Federal Register*. Public meetings may be held to discuss proposed research projects that will develop or evaluate products (versus policies or procedures) that have broad stakeholder vested interest and/or are potentially controversial.[3] The TI Research Program also organizes and hosts periodic National Occupational Injury Research Symposia (NOIRS), which bring together researchers from a broad range of disciplines to discuss research in progress and to form research and prevention partnerships. Workers, advocates, and other nonresearch groups may also attend and have an opportunity to provide input regarding TI research needs. The symposia is one means of implementing NORA for traumatic injuries. The TI Research Program has held four NOIRS since 1997 (the most recent in October 2008). At the inception of the four TI research programs currently directed by congressional initiatives and mandates (agricultural injuries among children, workplace violence, firefighter safety, and workers in Alaska's high-risk industries), NIOSH held stakeholder meetings to get input on possible research directions within these areas. TI Research Program staff also receive input through memberships with standards-setting and professional association committees (NIOSH, 2007a, p. 44) and the Council of State and

[3]Personal communication from N. Stout.

Territorial Epidemiologists. In addition to facilitating research project planning, engagement with stakeholders is an important means of identifying emerging TI research needs.

Identification of Emerging Research Needs

The TI Research Program uses surveillance data on fatal and nonfatal injuries—primarily from the Bureau of Labor Statistics (BLS) surveillance systems—to identify emerging research needs. The program also has real-time access to data on injuries reported at hospital emergency departments through the National Electronic Injury Surveillance System (NEISS); data on fatal injuries in selected states through the NIOSH Fatality Assessment and Control Evaluation (FACE) Program, which allow for quicker detection of injury clusters and spikes (NIOSH, 2007a, p. 46); and sentinel fatalities that identify previously unrecognized hazards.

Identification of emerging research needs also comes from TI Research Program engagement with stakeholders and partners. The series of NOIRS organized by the TI Research Program provide staff the opportunity to learn of emerging issues in traumatic injury research as well as adapt partnerships to address them (NIOSH, 2007a, p. 46). Identification of emerging problem areas in workplace trauma is one of the specific agenda items addressed at each of the NOIRS (NIOSH, 2007a,b). The NORA process and the relationships with stakeholders that result from it are also an important means of identifying emerging issues. In NORA I there was a committee that specifically addressed TI research needs; in the current NORA II process, industry-sector councils have been formed to address sector-specific research needs. Traumatic injuries are addressed by each of these councils.

The evidence package also describes features of the TI Research Program that facilitate response to emerging issues. These include the annual realignment of research priorities through the TI Research Program planning process, review of surveillance data and connectivity with stakeholders, and a flexible fatality investigation (FACE) program that allows for changes in the types of fatal incidents that are investigated (NIOSH, 2007a, p. 46). Notably absent is any reference to the Occupational Safety and Health (OSHA) regulatory agenda as input to program planning.

COMMITTEE ASSESSMENT OF THE TI RESEARCH PROGRAM'S PROCESS FOR TARGETING NEW RESEARCH AND IDENTIFYING EMERGING RESEARCH NEEDS

As discussed in more detail in Chapter 4, the committee is sensitive to the need for the TI Research Program to choose its research activities carefully to make the best use of limited resources.

Although the committee recognizes that both the severity and the magnitude of injury are important to consider when setting TI research priorities, it is concerned that the balance of focus between fatal and nonfatal injuries is either not evident or not optimal. Based on information provided in the evidence package, it is apparent that occupational fatality surveillance data have been an important program driver. Nonfatal occupational injuries far outnumber fatal injuries, and risk factors for fatal and nonfatal injuries are not necessarily the same. As was discussed in Chapter 2, while a focus on fatalities is reasonable in light of limited resources, it leaves a very substantial gap with respect to nonfatal injuries. The TI Research Program should be explicit about its consideration of fatal versus nonfatal data and provide a clear rationale when proposing and approving projects. Additional surveillance and surveillance research are needed to improve the characterization of nonfatal injuries (see Chapter 4 for a discussion).

The committee sees coordination of intramural and extramural research activities as essential to the TI Research Program's project planning process, especially given its limited resources and the importance of injury prevention and control research (see Chapter 4 for a discussion). Although the TI Research Program has opportunities for sharing of research ideas through several means—including NOIRS and staff participation on professional committees—the extent to which intramural researchers and extramural researchers are aware of each other's research and possible collaboration opportunities is not clear from the evidence package. Without some coordination of research activities, researchers may miss opportunities to access expertise as well as avenues for dissemination of research findings.

Stakeholder input into the TI Research Program's project planning and identification of emerging issues is important for ensuring the relevance of the research ultimately pursued. The committee sees NOIRS as important mechanisms for the TI Research Program to implement NORA and to foster the relevance of its research though dissemination. Alternative means of gathering external input may be necessary in years when NOIRS are not held. Research program external review sometimes includes a public meeting announced in the *Federal Register*. Additional strategies for gaining input from nonresearch groups—such as employers and employees—for both research project planning and identification of emerging issues should be considered as well.

Because the FACE Program collects real-time data on fatal occupational injuries, it allows for timely identification of previously unrecognized hazards. FACE investigations provide in-depth information about factors leading to fatal incidents that is not available elsewhere. This information is useful not only in generating hypotheses for future research, but also in developing recommendations for immediate prevention measures. FACE reports, which are based on these investigations and include recommendations to prevent similar incidents, are an important vehicle for reaching employers, workers, and product manufacturers with prevention messages. Investigations conducted by the NIOSH FACE Program and state-based FACE programs have facilitated the development of numerous interventions, including changes in workplace policies and practice, as well as equipment design. In conjunction with population-based fatality statistics, FACE data can assist NIOSH and the TI Research Program with strategic selection of problems upon which to focus future research.

COMMITTEE REVIEW OF THE TI RESEARCH PROGRAM'S STRATEGIC GOALS FOR THE FUTURE

Following the 2005 implementation of a matrix management structure to coordinate cross-institute programmatic activities, a NIOSH TI steering committee was formed to develop strategic goals and outcome measures for NIOSH TI research. The steering committee, with input from staff, developed a list of TI strategic goals for the future. The evaluation committee was informed that the list of goals is evolving as NIOSH considers external input. In the evidence package, the evaluation committee was invited to comment on the TI Research Program's strategic plan to help "maximize the relevance, quality and impact of the NIOSH TI Research Program" (NIOSH, 2007d, p. 195).

The TI Research Program's five strategic goals for the future are to (1) reduce fall injuries in the workplace; (2) reduce occupational injuries and deaths due to motor vehicles; (3) reduce occupational injuries and deaths due to workplace violence; (4) reduce occupational injuries and deaths due to machines and industrial vehicles; and (5) reduce occupational injuries and deaths among high-risk and vulnerable worker groups.[4] Within each of the five strategic goals are three or four subgoals that generally identify types of injuries, worker populations, industries, and workplace exposures on which to focus. At the next level are intermediate goals for achieving goal and subgoal objectives (see Box 3-1 for the goals and subgoals. The complete list with intermediate goals can be found in Appendix D).

The following section outlines the committee's review of the TI Research Program's proposed research within the five strategic goal areas. Based on member expertise, the committee identified areas of research that warrant attention in the future. Some of these are described in the context of the review of the strategic goals, and some are described in a subsequent section. Given limited staffing and budget resources, it is not expected that the TI Research Program will pursue all of the proposed research areas, but rather that it will take them into consideration.

---

[4]The numbering of the goals here is consistent with the numbering of the goals as presented in the evidence package prepared by NIOSH for the committee. Numbering is not a ranking of goals by research priority.

BOX 3-1
TI Research Program Strategic Goals for the Future

1. **Reduce fall injuries in the workplace**
   1.1. Reduce fall-related fatalities and injuries in the construction industry
   1.2. Reduce fall-related injuries in the health services industry
   1.3. Reduce fall-related injuries in the wholesale and retail trade industry
   1.4. Reduce fall-related injuries through research on biosciences underlying human fall initiation, fall dynamics, fall termination, and control measures
2. **Reduce occupational injuries and deaths due to motor vehicles**
   2.1. Reduce motor vehicle-related incidents in the transportation, warehousing, and utilities (TWU) industry sector
   2.2. Reduce fatal and serious nonfatal injuries to workers in roadway construction work zones
   2.3. Reduce injuries and fatalities from motor vehicle incidents, including being struck by vehicles, among public safety and emergency response workers
   2.4. Reduce occupational road traffic injuries worldwide
3. **Reduce occupational injuries and deaths due to workplace violence**
   3.1. Reduce workplace violence in the transportation, warehouse, and utilities industries
   3.2. Reduce workplace violence among high-risk wholesale and retail trade workers including grocery stores, gasoline stations, convenience stores, bakeries, and liquor stores
   3.3. Identify risk factors and effective interventions to prevent workplace violence among high-risk services, healthcare, and social service sector workers such as eating and drinking establishment workers; hotel or motel workers; automotive repair mechanics; teachers; nurses and nursing assistants in general medical, home health care, nursing homes, and psychiatric hospitals; social service workers in job training, residential care, and day care industries; private security workers; and public safety and correctional workers in emergency response tasks (e.g., medical services, police calls, and correctional officer activities)
4. **Reduce occupational injuries and deaths due to machines and industrial vehicles**
   4.1. Reduce occupational injuries and deaths due to machines and industrial vehicles in the agriculture, forestry, and fishing industry, with an emphasis on tractor-related injuries and deaths
   4.2. Reduce occupational injuries and deaths due to machines and industrial vehicles in the construction industry
   4.3. Reduce occupational injuries and deaths due to machines and industrial vehicles in the manufacturing industry
   4.4. Reduce occupational injuries and deaths due to machines and industrial vehicles in the mining industry
5. **Reduce occupational injuries and deaths among high-risk and vulnerable worker groups**
   5.1. Reduce occupational injuries and deaths among young workers
   5.2. Reduce occupational injuries and deaths among older workers
   5.3. Reduce occupational injuries and deaths among high-risk ethnic and minority workers
   5.4. Reduce occupational injuries and deaths among immigrant workers

Evaluation Committee's Assessment of TI Strategic Goals for the Future

The committee's comments on each of the TI Research Program's five strategic goals areas are given below. Overall, the committee finds that the TI Research Program's strategic goals are focused on major contributors to occupational injuries and deaths and are sensitive to populations and groups at disproportionate risk. Among the strategic goals are several intermediate goals for leveraging partnerships that, if carried out, could help the TI Research Program a great deal in maximizing impact.

In certain areas (specifically, in the future goals for motor vehicles), in future iterations of its strategic goals the TI Research Program should work toward focusing its efforts on areas that are not researched by other agencies or covered by other agency missions. Future goals should be evaluated and updated regularly. In general, the committee also feels that future modifications to the goals could include better indication of how proposed interventions and partnerships will be evaluated.

Strategic Goal 1: Reduce Fall Injuries in the Workplace

The TI Research Program strategic goal for fall injuries addresses an important source of fatalities from occupational injuries. The committee finds the planned efforts targeting falls from elevations in the construction industry (intermediate goals 1.1.1 and 1.1.2) to be especially appropriate given the high frequency of fatalities from falls in that industry.

One of the weaknesses of the current program described in Chapter 2 is the lack of formal recognition by the TI Research Program of slips, trips, and falls (STFs) at the same level. The strategic goals also reveal little planned research on same-level STFs for the future. Intermediate goal 1.2.2 is to identify and summarize research on STFs applicable to the health services industry, and same-level falls may be an implicit focus of this goal. Same-level STFs are a major source of work-related morbidity. The committee believes it would be relevant for the TI Research Program to make surveillance of same-level STFs among health services and other workers a focus for future research.

With respect to STF risk factor and intervention research, one of the major issues continues to be the lack of understanding of tribology for both wet and dry surfaces and the effects of the various polymers used in shoe materials. There is currently no available means of measuring slipperiness as opposed to the various measures of the coefficient of friction. The TI Research Program could initiate progress in this area, perhaps through collaborations with other agencies such as the National Science Foundation, which has a tribology section.

One justification for expansion of the falls program to include additional research on STFs at the same level is the increasing number of people working beyond the normal age of retirement who are at higher risk of injury from falls. In a presentation before the committee, the TI Research Program identified an aging workforce as an emerging issue in fall prevention research. There could be crosscutting synergies between research on same-level falls and the TI Research Program's planned goals for reducing occupational injuries and deaths among older workers (see strategic goal 5.2).

Strategic Goal 2: Reduce Occupational Injuries and Deaths Due to Motor Vehicles

Transportation incidents are consistently the leading cause of fatal occupational injuries. The committee therefore considers it appropriate for the TI Research Program to continue work in this area where there is a specific occupational context and an opportunity to make a unique contribution. Increased collaborations with the National Highway Traffic Safety Administration (NHTSA) and other Department of Transportation agencies will be key to increasing the profile of the work-relatedness of motor-vehicle injuries.

Strategic subgoals 2.2 and 2.3 for the prevention of motor vehicle injuries in roadway construction work zones and among public safety and emergency response workers address important contributors to occupational motor vehicle-related traumatic injury and death and are relevant for the TI Research Program. In the area of construction work zones, the committee in particular supports the planned continued efforts to evaluate the proximity warning systems (PWS) and internal traffic control plan (ITCP) interventions for use on road construction sites (intermediate goals 2.2.2 and 2.2.3). The committee also supports the planned continued research on some motor vehicle-related causes of occupational injuries among emergency responders (subgoal 2.3) and the promulgation of information on safety restraints.

Given its limited resources and focus on occupational injuries, the committee found subgoal 2.1—to reduce motor-vehicle-related incidents in the transportation, warehousing, and utilities (TWU) industry sector—to be rather broad for the TI Research Program to pursue. The committee sees this goal and its intermediate goals 2.1.1, 2.1.3, and 2.1.4 as being more appropriately addressed by an organization such as NHTSA, which has more resources and a focus on general highway safety rather than occupational safety. As discussed below, however, the committee does believe that the TI Research Program could contribute to research and intervention on the occupational aspects of short-haul trucking.

Subgoal 2.4—to reduce occupational road traffic injuries worldwide—is a strategic goal of the NIOSH Global Collaborations Program in TWU. Efforts of this program are intended to benefit both U.S. workers and workers globally through international partnerships and information sharing (NIOSH, 2008). The committee finds this subgoal relevant for the TI Research Program. Indeed, the United States has much to learn from other countries in the area of occupational safety and health. In the area of motor vehicle research in particular, as stated elsewhere, NIOSH should be quite focused and judicious in the allocation of limited resources. International research then should promote benchmarking (learning from those with better practices) and the focused motor vehicle goals established (e.g., work zone safety).

Gaps in motor vehicle injury research identified by the committee include taxi driving, short-haul trucking, day delivery drivers, parking lot occupational driving, and intrastate driving. Focusing on taxis should offer crosscutting synergies within NIOSH; taxi driving is a hazardous occupation because of both workplace violence and the occupational driving itself. Taxi driving creates risk not only for the drivers, but also for the occupants of the taxi, pedestrians, and occupants of other vehicles. While much research is done on long-haul trucking safety by agencies such as NHTSA, traumatic injuries associated with short-haul trucking (defined as trips of less than 50 miles) are underserved in terms of research and intervention. Short-haul trucking accounts for the majority of the trucking industry (Hanowski et al., 2000). In addition, there may be an opportunity to address the nondriving aspects commonly involved in short-haul trucking such as lifting and handling that are not as common for long-haul trucking. Day delivery driving is a concern because of both the driving aspects and package delivery aspects of the job (Hira, 2007). Slips and falls, lower back injury, and sprains and strains are particular concerns. Other areas of potential opportunity for high-impact research include parking lot and rest area traumatic injuries related to short-haul trucking and intrastate driving.

Strategic Goal 3: Reduce Occupational Injuries and Deaths Due to Workplace Violence

The activities outlined in the future goals for workplace violence demonstrate good research program transitioning—beginning with surveillance and identification of risk factors—and moving into design and implementation of interventions where needed. The committee finds these goals to be a good model for how to build on previous achievements in the other goal areas. Workplace violence goals also appear to be well aligned with NORA II, with objectives focused on research in the transportation, healthcare, service, and retail sectors.

Although its goals are generally directed appropriately toward understanding the most high-risk situations, the TI Research Program might consider focusing in subgoal 3.2 on workplace violence against retail workers only, rather than against both retail and wholesale workers, because more violence occurs in retail than in other workplace environments. The committee finds intermediate goals for the development of partnerships with organizations, associations, police departments, and employers to be appropriate for maximizing research relevance and impact, but a corresponding evaluation component is needed to assess the effectiveness of partnerships.

Worker-on-worker (or Type III) workplace violence and domestic violence in the workplace (which is included in Type IV) appear to be absent from the strategic goals, although they may be included implicitly. The committee finds study of worker-on-worker violence (e.g., bullying) and domestic violence in the workplace to be important areas of research for NIOSH, but they may be best addressed within the NIOSH Work Organization and Stress-Related Disorders Program, rather than the TI Research Program.

Strategic Goal 4: Reduce Occupational Injuries and Deaths Due to Machines and Industrial Vehicles

This strategic goal includes plans to complete testing and market development of the auto-deploying rollover protective structure (AutoROPS) for use in low-clearance farming and the cost-effective ROPS (CROPS) for adoption by manufacturers (see intermediate subgoals 4.1.1-4.1.3). These subgoals represent a continuation of previous efforts by the TI Research Program to address barriers to the use of ROPS on tractors by farm workers. The work of the TI Research Program on ROPS is to be commended, and the committee supports its future goals for AutoROPS and CROPS. However, studies (Hallman, 2005; Spielholz et al., 2006) indicate that retrofitting of older tractors is occurring at a slower than optimal pace and that, in addition to cost, a "hassle factor" has been an obstacle to retrofitting even when there is a financial incentive for doing so. The TI Research Program could devote more resources toward technology adoption research to facilitate faster farm worker adoption of this important technology.[5]

In addition to continued work on AutoROPS and CROPS, the strategic goals include plans to develop and/or evaluate several other technologies to prevent machine-related injuries, such as the machine emergency-stop (or "e-stop") system for use on commercial fishing vessel machinery (intermediate goal 4.1.6) and a

[5]Although not indicated in the evidence package, the committee learned that, in 2006, NIOSH provided funding to several university-based agricultural safety and health research centers to explore techniques to promote safe use of tractors. Under the initiative, the centers are studying some of the barriers to the use of ROPS on tractors, such as financial incentives to retrofit tractors with ROPS and the impact of changes to standards, regulations, and technology, and their effect on future ROPS availability. Several of the centers are also testing community-based social marketing techniques to improve safe use of tractors (NIOSH, 2006).

proximity warning system for use on road construction equipment (intermediate goal 4.2.2). If widely adopted, these technologies are likely to have a great impact on reducing machine-related injuries. Evaluation of the protective technologies themselves, as well as the barriers to their adoption, will be important for successful implementation.

A gap in the TI Research Program's planned future research on machine-related injuries is the prevention of injuries resulting from devices used for landscaping, such as riding and zero-turn lawn mowers, mechanical hedge trimmers, cultivators, and others. Landscaping and horticultural work has been identified as one of the most hazardous industries, and machinery is one of the sources of injury within that industry.

Strategic Goal 5: Reduce Occupational Injuries and Deaths Among High-Risk and Vulnerable Worker Groups

The U.S. workforce is aging and becoming increasingly racially and ethnically diverse. Recognizing these changing demographics of the U.S. workforce, the committee supports the addition of research on older workers and immigrant and minority workers to the TI Research Program's research goals for the future. The committee also supports the continuation of research on youth, because this group continues to have unique research needs. The TI Research Program should consider expanding its current focus on workers younger than age 18 to the 18-24-year-old age group, which has the highest nonfatal occupational injury rate. The next decade will witness an influx of new workers in this age range as the children of the baby boomer generation enter the workforce in peak numbers. (See section on The Changing Nature of Work in Chapter 4 of this report.)

The strategic goals for high-risk and vulnerable groups include improved surveillance (intermediate goals 5.1.1, 5.2.1, 5.3.1, 5.4.1). Innovative approaches will be necessary to document the experience of vulnerable worker groups who are not well captured in the conventional occupational health data sources. The strategic goals (intermediate goals 5.2.2, 5.3.2, 5.4.2) include identification of risk factors for injury and death among older workers and among minority and immigrant workers. The committee feels there is also a continued need for risk factor research on young workers. It will be important to extend risk factor research on vulnerable workers to include the study of informal and formal workplace policies and workplace norms, and alternative work arrangements, as well as individual characteristics that may contribute to disparate risks. Research on individual worker characteristics should include the exploration of social and cultural as well as the more commonly recognized language barriers. In its review of the TI Research Program's current goals the committee saw little evidence that research on vulnerable worker groups had been a high priority, and program efforts addressing immigrant and minority workers seemed to focus predominately on language barriers.

The TI Research Program strategic goals include a specific intermediate goal (5.2.3) to evaluate intervention strategies to prevent older worker injuries and deaths. The committee sees a need for evaluation of intervention strategies for preventing injuries and deaths among youth, minority workers, and immigrant workers as well. Community-based participatory approaches should be included. Collaboration with the NIOSH Occupational Health Disparities program will be important to improve research capacity in this area. While the committee supports plans for partnerships (intermediate subgoals 5.2.4, 5.3.3, 5.3.4, and 5.4.3) with government agencies, safety groups, and other organizations, the TI Research Program might consider adding an evaluation component to measure how well the partnerships are working.

Two populations that are not specified in the TI Research Program's strategic goals and for which research is currently lacking are workers with developmental or physical disabilities and workers in nontraditional work arrangements (e.g., contract workers, day laborers). Little research has been performed either to track injuries among people with disabilities or to identify their unique risk factors for injury. Intervention research is needed for workers in nontraditional work arrangements that may involve highly hazardous work. Because of lack of oversight, inadequate training, lack of access to health and safety resources, and other factors, these workers may not be afforded the same protections as permanent workers.

One area of current TI Research Program research that is not reflected in the strategic goals for the future is research on acute back injuries. According to TI Research Program staff,[6] work on acute back injury will be moved out of the TI Research Program and into the NIOSH Musculoskeletal Disorders Research Program. The committee supports NIOSH's continuation of research on back injuries, which continue to be among the most serious and costly occupational health problems.

The committee learned from senior TI Research Program staff that while there is no longer an Alaska-specific goal in the proposed goals for the future, the TI Research Program intends that the work of the Alaska Field Station (recently renamed the Alaska Pacific Regional Office) will continue to contribute to the cause-specific goals of the TI strategic plan. Intermediate goals 4.1.4-4.1.8 for machinery are a continuation of some of the traumatic injury research being performed under the current goal to reduce injuries and fatalities among workers in Alaska's fishing and logging industries.[7] The committee finds that NIOSH's research in Alaska to date

[6]Personal communication from N. Stout.

[7]Personal communication from N. Stout.

has been impressive and can serve as a model for moving research to practice in other states and in other goal areas.

The TI Research Program will continue research on emergency responders within its proposed strategic goals for motor vehicle-related injuries (subgoal 2.3). The committee supports risk factor and intervention research for this population, as planned in intermediate goals 2.3.1 and 2.3.2. Through past investigations of firefighter and emergency medical services (EMS) worker deaths in ambulance crashes, the TI Research Program has demonstrated a need for better passenger protective equipment (NIOSH, 2007c, p. 150) and has been investigating engineering interventions. Plans to develop and promulgate information on restraints to firefighter and EMS workers (intermediate goal 2.3.3) are an appropriate extension of this research.

Other Gaps in Traumatic Injury Research

In addition to gaps in research noted within the context of the five strategic goals, the committee identified other priority research areas that the TI Research Program could pursue in the future. Without attempting to convey a complete roster of additional research needs, the committee offers five illustrative areas that warrant attention in the future:

1. *Organizational culture and adoption of safety measures.* Successful translation of safety research into practice is dependent upon the culture—including the attitudes, experiences, core beliefs, and values—of a workplace. Instituting measures to protect worker safety and health can be difficult when such measures are inconsistent with elements of an organization's culture. In a workplace where there is high emphasis on short-term production (rather than longer-term productivity), for example, employees may be less likely to make use of a safety measure they feel might slow them down, because of the competing demand to produce more in a given amount of time. While culture is often acknowledged as an important factor inside and outside NIOSH, translation and dissemination of research into the many elements of organizational culture that facilitate safety prevention and inhibit employer adoption of safety measures are needed.
2. *Cost of injuries.* When safety measures are being considered by an employer, they often must compete with carefully detailed cases for other employer investments (such as equipment and facilities) that are usually supported by a statement of the cost of implementation and a prediction of improvements in productivity or reduction of labor measured in

dollars. These estimates allow a rational, benefit-cost ration-based decision to be made by the employer. Comparable benefit-cost estimates for adoption of safety measures are often not possible. The hidden indirect costs that an employer might save by adopting a safety measure, such as the costs associated with employee time away from work due to an injury or the costs of training new employees, can be difficult to estimate. There is a need for research to develop both an agreed method of calculating benefits for adoption of safety measures and an updated publication of the costs of various injuries.

3. *Policy evaluation research.* The committee sees a role for the TI Research Program in the evaluation of both public and company policies that impact the occurrence of occupational injuries. For example, as an extension of plans to provide the empirical data to guide the development and enforcement of child labor laws (intermediate goal 5.1.3), the TI Research Program could also be involved in the evaluation of such laws if they are enacted. Other important policy areas to evaluate might include the effectiveness of state requirements for maintaining workplace safety programs; workplace health and safety committees required by state laws or collective bargaining agreements; and cost of compliance or noncompliance with specific OSHA regulations (e.g., fall protection, trenching, ladders).
4. *Small- to medium-sized enterprises (SMEs).* SMEs seem to be particularly vulnerable to the perceived trade-off between safety and productivity and other cultural issues mentioned earlier. In addition, there are many workers in potentially dangerous occupations that work for small companies and contractors that are unlicensed, not certified, and have no safety programs. In many of these settings there is a large representation of immigrants, minorities, and other vulnerable groups. Surveillance research is needed to determine the extent and types of injuries that occur in SMEs, particularly nonfatal injuries for which reporting has been a problem. Findings from such research would help provide the basis for determination of whether and how injuries that occur in SMEs differ from those that occur in larger enterprises and could inform SME-targeted risk factor and cost-sensitive intervention research as well as evidence-based best practices for SMEs that seek to reduce their work-related risks and costs. It has been well documented that small businesses face special challenges in complying with safety regulations and safe work practices. These challenges include lack of access to information, lack of access to expertise, and lack of capital for investment in safety. Special attention is needed to research the

barriers to the adoption of safety measures by small businesses, and the incentives and resources needed to overcome these barriers.

5. *Surveillance.* The success of the TI Research Program depends on a robust surveillance system. As indicated in Chapter 2 and as will be discussed in Chapter 4, the committee believes that NIOSH would benefit from an overall strategy regarding surveillance research and implementation. Improving the surveillance system and surveillance research is of sufficient importance to warrant inclusion as a separate strategic goal for the TI Research Program.

Stakeholder Comment

As part of its evaluation, the committee sought the input of stakeholders on the relevance and impact of the TI Research Program's current research portfolio. Responses to this request included several suggestions for additional research. A summary of these suggestions, along with an outline of the stakeholder comment request process, can be found in Appendix B of this report.

REFERENCES

Hallman, E. M. 2005. ROPS retrofitting: Measuring effectiveness of incentives and uncovering inherent barriers to success. *Journal of Agricultural Safety and Health* 11(1):75-84.

Hanowski, R. J., W. W. Wierwille, S. A. Garness, T. A. Dingus, R. R. Knipling, and R. J. Carroll. 2000. A field evaluation of safety issues in local/short haul trucking. In *Proceedings of the XIVth Triennial Congress of the International Ergonomics Association and 44th Annual Meeting of the Human Factors and Ergonomics Society*, Santa Monica, CA.

Hira, N. A. 2007. The making of a UPS driver. *Fortune*, November 7.

NIOSH (National Institute for Occupational Safety and Health). 1998. *NORA: Traumatic occupational injury research needs and priorities.* Washington, DC: HHS.

NIOSH. 2006. NIOSH Awards Funding to Ag Research Centers for Initiative to Reduce Tractor Deaths, Injuries. NIOSH Update. March 21, 2006. http://www.cdc.gov/niosh/updates/upd-03-21-06.html (accessed July 24, 2008).

NIOSH. 2007a (unpublished). *Overview of the TI Research Program.* In the evidence package provided to the Committee to Review the NIOSH TI Research Program. NIOSH.

NIOSH. 2007b. *2008 national occupational injury research symposium.* http://www.cdc.gov/niosh/noirs/2008/about2008.html (accessed November 20, 2007).

NIOSH. 2007c (unpublished). *Current TI research goals and sub goals: Reduce injuries and fatalities to emergency responders.* In the evidence package provided to the Committee to Review the NIOSH TI Research Program. NIOSH.

NIOSH. 2007d (unpublished). *TI strategic goals for the future.* NIOSH.

NIOSH. 2008. NIOSH program portfolio: Global collaborations in transportation, warehousing and utilities. http://www.cdc.gov/niosh/programs/twu/global/ (accessed July 24, 2008).

Spielholz, P., T. Sjostrom, R. E. Clark, and D. A. Adams. 2006. A survey of tractors and rollover protective structures in Washington State. *Journal of Agricultural Safety and Health* 12(4):325-333.

4

Recommendations for Program Improvement

STRATEGIC PLANNING

As described in Chapter 1, the current goal areas represent a retrofitting of a decade's worth of work into a structure to be reviewed with the Framework Document in mind. This does not mean that considerable thought was not given by the National Institute for Safety and Health's (NIOSH's) Traumatic Injury (TI) Research Program over the decade to the work being done, but from the presentation of the goals to the committee in the evidence package, it is clear that some of the TI Research Program's effort occurred outside and independent of a program-wide coherent planning process.

The committee recognizes that opportunities arise and an agency must be adroit to deal with unexpected events (for example, the work in goal 2 regarding falls from telecommunications towers appears to be a response to a newly discovered occupational risk) and that there are points of departure from any planning document. Some work will appear not to fit in well with the rest of the program. This is not unexpected. With some exceptions, as discussed in Chapter 2, the program worked in areas of public health importance and documented intermediate outcomes.

The committee has concluded that the TI Research Program successes (as defined by the activities, outputs, and outcomes reviewed) occurred most obviously in goal areas in which there was a focused and intense effort (due to resources or a geographic focus, e.g., Alaska) or in which a goal was narrowly defined or clearly defined and achievements could be documented. Often these areas of focus were due to congressional directives, increased resources associated with congressional directives, and staff interest and expertise. Some of the successes arose from a focus on a newly emerging concern.

The TI Research Program should be careful in its next stage of planning and priority setting to outline as specifically as possible the scope of the work it plans to accomplish so that its achievements are demonstrably linked to a problem of importance and their research is strategic. The committee urges the TI Research Program to focus on those occupational risks that it has specific skills for addressing and that are not currently or better addressed by other federal or nonfederal researchers. Otherwise, it risks competing with other agencies and squandering precious resources on activities that could be redundant or that will not necessarily or directly lead to accomplishing the goals of reducing morbidity and mortality from occupational traumatic injuries.

The committee does not mean that the TI Research Program should *only* work in areas that might be seen as "low-hanging fruit" or that can easily be shown to be impacted quickly. Some work will take a long time to show results, especially intermediate outcomes, and this is true for research in any field. The TI Research Program will likely need to address a mixture of goals that have potential for short-term and long-term impact; explicitly outlining a time line for expected results will make the evaluation of success more obvious.

An important component of strategic planning is to consider criteria for ending work in a specific goal or ending a specific approach to achieving the goal. As described in Chapter 2, the committee felt that the work in some goals lasted beyond its period of usefulness to the detriment of possibilities to impact other important TI problems. For example, the work on subgoal 4.1, the rollover protective structure (ROPS) program, seemed to reflect a missed opportunity to move beyond the "comfort zone" of the team working in this area. The committee (and other evaluation committee reviews) applauded the earlier work in ROPS but felt that persevering in developing engineering solutions occurred at the expense of a plan to explore other means, such as policy solutions, to overcome the barriers to more widespread use of ROPS.

1. **Continue setting goals that are within the TI Research Program's scope and resources.** Given its limited resources, the TI Research Program should continue a research focus and priority setting on goals that are well defined, are based on rigorous surveillance data, and are complementary to work being done by stakeholders, extramural research partners, or other agencies.

2. **Develop an explicit plan for each subgoal.** The TI Research Program should develop an explicit, written plan within each subgoal for progression along the public health framework, including the circumstances under which work in the subgoal should cease. Additional considerations should be the relative balance between risk factor and intervention research.

COORDINATION AND COLLABORATION

Given that the TI Research Program operates under severely limited resources, it not only must be strategic in selecting its priorities, as discussed previously, but must also position itself to benefit from collaborations within the federal government and with academic researchers and state agencies. These collaborations and coordinating activities will help the TI Research Program prioritize its activities in order to complement work elsewhere or to avoid duplication of effort. Coordination and collaboration can be achieved by several means, such as organizational relationships and research programs. As such, the committee offers several recommendations that are intended to support and encourage some current collaborations, to identify a major new initiative, and to leverage a relationship that has perhaps recently been ignored.

NIOSH is one of several federal agencies with a role in injury prevention and control, so there is obviously some overlap of agency interests, particularly with regard to research and information dissemination. In focusing on occupational traumatic injury, NIOSH should continue to foster and build relationships with other agencies, (e.g., the National Highway Traffic Safety Administration, the National Institutes of Health [NIH], the National Center for Injury Control and Prevention, the Consumer Product Safety Commission, the Department of Defense, the Health Resources and Services Administration, and the Agency for Healthcare Research and Quality). This is not to suggest an absence of communication among agencies. The committee encourages continued and appropriate attention to interagency issues in order to assure a unique research portfolio in the TI Research Program and the efficient deployment of scarce resources.

3. **Work with other federal agencies that support injury prevention and control research.** NIOSH and its TI Research Program should work with senior leadership from other federal agencies to outline areas of collaboration and synergy; to identify opportunities to further the science of injury control and prevention; and to reduce the burden of injury across populations, environments, and products.

As discussed in Chapters 2 and 3, NIOSH played a very important role in developing surveillance of fatal occupational injuries. Together with the Fatality Assessment and Control Evaluation Program (FACE; case investigations that are conducted by TI Research staff or through state-based programs), this has allowed the TI Research Program to better understand the burden of injury, risk factors for injury, progress toward decreasing occupational injuries, and program prioritization. However, the program has clearly been driven by a focus on fatalities. The committee understands that there are legitimate reasons to focus on fatalities and legitimate debates about aggregate burdens, but fatal injuries are not proxies for nonfatal injuries. Risk factors can vary widely. The time has come for the TI Research Program to bolster its focus—particularly starting with population-based surveillance—on nonfatal occupational injuries without lessening the excellent work on fatal injuries. This will require work with other agencies, including the Occupational Safety and Health Administration (OSHA) to improve the usefulness of the worker logs and with the Bureau of Labor Statistics to improve the Annual Survey of Injuries and Illnesses and to better understanding the reasons for underreporting. The TI Research Program could consider extending the FACE Program to specific nonfatal injuries.

4. **Improve surveillance of nonfatal injuries.** The TI Research Program should develop a plan for improving surveillance of nonfatal injuries, integral to prevention and to strengthening the TI Research Program portfolio development. A comprehensive approach should go beyond use of employer-based data to include nonemployer-based data sources such as hospital data and other medical data systems, the National Health Interview Survey, and the Behavioral Risk Factor Surveillance System. The TI Research Program should involve other relevant federal and state agencies in developing a cohesive interagency effort.

As noted in Chapter 2 as well as other evaluation committee reports, the role of OSHA in implementing regulatory actions is a significant external factor to the success of NIOSH research. The committee understands the difficulty NIOSH might experience in this regard. OSHA is organizationally distinct by statute (the rationale for this is not being questioned in this report). The committee urges perseverance on the part of the TI Research Program in addressing injuries and interventions that are amenable to regulatory action by or otherwise of interest to OSHA. It should also increase its engagement with the Authoritative Recommendation Program at NIOSH to promote translation of TI Research Program research findings into OSHA regulations and consensus standards.

5. **Work collaboratively with OSHA.** An agency of particular importance and relevance to NIOSH is OSHA. The TI Research Program, along with NIOSH leadership, should continue to work with OSHA to identify areas of high-priority research that NIOSH should undertake and to identify NIOSH research findings of particular salience for potential regulatory action by OSHA.

As discussed in Chapters 1 and 2, NIOSH is an agency that supports intramural and extramural research. Its extramural partners include state occupational public health programs, which conduct surveillance, FACE investigations, intervention research, training, and a range of transfer and dissemination activities. Managing the appropriate sets of responsibilities could occasionally prove challenging, because each partner has specific interests, needs, and other funding opportunities and responsibilities.

The committee fully understands the tension between the needs of federal agencies and the interests of academic researchers in particular. Academic researchers are not necessarily interested in the pragmatic questions that federal agencies need addressed. Academic researchers also tend to enjoy a degree of freedom in research settings that might not be compatible with the data or research requirements of an agency with a very narrow and directed mission. Conversely, agency priority-setting activities might not allow for sufficient time for nondirected basic research in the academic setting to show relevance to agency needs. Finally, agencies often need results disseminated promptly, which is not always compatible with the traditional publication process. There is a spectrum of approaches to this dilemma—ranging from structuring the intramural program to fill gaps not likely to be filled by extramural researchers, to directing extramural research funding to fill gaps within the intramural program, as well as everything in between. The committee encourages the TI Research Program to follow a middle ground. That is, the intramural program has obvious areas of expertise that should be used to the fullest, while new hires could be used to expand the intramural capability. The extramural research program can "fill in" what the intramural program lacks, as well as provide innovative approaches not currently anticipated or realized.

The TI Research Program has research facilities that, although not always unique, are well established and productive. For example, the anthropometric labs and the virtual reality lab are excellent, and intramural research should use them as much as possible. Making these facilities available to extramural partners could expand their usefulness.

Extremely large research-granting programs might find it difficult to keep track of all the extramural research activities and outcomes, but the TI Research Program is small enough that, with good strategies in place, it should be able to structure intramural research projects that complement, not compete with, academic partners. Similarly, extramural research opportunities that complement staff expertise can be targeted. Picking the most strategic mix of intramural and extramural research does require reporting from extramural researchers, which can seem burdensome to them. However, with its limited research budget, the TI Research Program deserves and needs to fully understand the work it is supporting. Communication and collaboration are key to a successful mix of intramural and extramural research. Increased communication can also lead to increased opportunities for appropriate transfer of research findings into practice.

The committee urges that this collaboration not be overly directed or programmed. Rather, the committee sees this as an important step in building teams for TI research. The committee recognizes that there are obvious benefits to increased interactions between researchers and NIOSH should facilitate such interactions rather than impede them, such as through bureaucratic barriers. Specific examples could include the use of NIOSH laboratories by academic researchers, co-publication of results from complementary research, and information exchange about the status of research projects. The committee is aware of regular, collaborative interactions between intramural and extramural researchers in other NIOSH programs and encourages the TI Research Program to explore those models if it has not done so already. Further, the committee urges that NIOSH extend its efforts in collaboration and coordination within NIOSH programs. The matrix management model, as discussed in Chapter 2, does not guarantee the kind of collaboration that it strives to foster.

6. **Ensure collaboration among NIOSH-funded researchers.** NIOSH should review its practices and take steps to improve the opportunities for intramural and extramural researchers, including state occupational public health programs, to communicate and collaborate without excessively directing extramural research to the detriment of scientific creativity. NIOSH should also further ensure collaboration and coordination among its programs, including the traumatic injury, construction, mining, and agriculture programs.

WORKFORCE DEVELOPMENT

Researchers in occupational injury claim that there are few academic programs to train students or postdoctoral fellows for pursuing a career in researching traumatic injury. As the health research field has moved increasingly toward molecular or very high technology tools, young students increasingly choose a more basic science career. Although biomedical funding (e.g., through NIH) has not increased the last few years as it did in the recent decade, it is certainly an attractive career direction for budding researchers. The more applied nature of injury research does not seem to grab the attention of students. There are currently few graduate programs in occupational injury research. NIOSH created a program for encouraging occupational research in universities through the Education Research Centers (ERCs) Program. NIOSH supports 17 ERCs in the core areas of industrial hygiene, occupational health nursing, occupational medicine, and occupational safety, plus specialized areas relevant to the occupational safety and health field. Most NIOSH-supported research at the ERCs, however, seems to be in the "health" component rather than the "safety" component. An important part of encouraging traumatic injury research at ERCs would be to expand the definition of "safety programs" and frame them more broadly to include occupational injury epidemiology and prevention, injury prevention intervention research, and other more forward-looking disciplines within the safety arena. ERCs could also expand their reach by increasing interactions with their institutions' epidemiology and engineering programs.

The TI Research Program needs to develop a plan to increase the pipeline of traumatic injury researchers. In order to do so, the committee suggests stressing the technical expertise required to work in traumatic injury, the interdisciplinary nature of successful traumatic injury research, and the possibility of "making a real difference" with traumatic injury research.[1]

7. **Increase the visibility of traumatic injury research.** NIOSH should embark on a program to increase the visibility of traumatic injury research in order to attract new researchers. Absent a significant increase in research funding, the TI Research Program can still attempt to influence the number of ERCs that have a focus on safety research and can still disseminate information about the quality, impact, and scientific challenges of traumatic injury research, as well as the dynamic changes in the field that go beyond the confines of traditional safety engineering.

TRANSFER

The most obvious federal partner for transferring traumatic injury research into practice is OSHA. However, as has been noted in other parts of this report, NIOSH has no control over OSHA regulatory decisions and the committee has addressed this in a recommendation made earlier. However, not all interventions

[1]This is consonant with advice from a recent report addressing the workforce pipeline challenges in engineering research (NAE, 2008), *Changing the Conversation: Messages for Improving Public Understanding of Engineering.*

fall into the purview of regulatory action and NIOSH works to reduce traumatic injuries in other ways. NIOSH created a research-to-practice (r2p) initiative with six components: prioritize, partner, target, translate, disseminate, and evaluate. This initiative recognizes the role of partners in this collaborative process. The TI Research Program has a long history of successful partnerships with industry and the committee hopes this will continue. The TI Research Program review in Chapter 2 includes several good examples of r2p efforts, most notably the work of the Alaska Field Station. However, the committee is concerned that the TI Research Program, and perhaps all of NIOSH, is not fully prepared to rigorously and expertly execute an r2p enterprise.

To improve on this initiative, it is important that the TI Research Program allow its talented staff and extramural researchers or other grantees to focus their efforts, play on their strengths, and collaborate with others to complement their own expertise. Experts in translation should be included in project teams. As the TI Research Program develops better tracking of extramural research projects, translation activities regarding the outcomes of this research can be planned, whether through translation components included in the extramural research or by collaborations with the NIOSH transfer experts. Talented basic scientists should not spend their time on transfer activities that are beyond their expertise. This is not to suggest that the basic scientists or engineers should be absolved from considering the transfer implications of their work; each project should involve a plan for transfer of the appropriate type at the appropriate time. The committee notes that it will be important for the TI Research Program to maintain a balance between basic research, applied research, and transfer activities. The committee further notes that NIOSH has no control over OSHA, its primary federal partner in transfer activities. The transfer of research to practice is important in its own right for nonregulatory improvements in injury prevention and control, but OSHA receptivity to regulatory action would make the impact of r2p even greater.

8. **Evaluate research-to-practice efforts.** NIOSH should develop a strategic plan for evaluating its r2p efforts and for building the capacity to carry out and evaluate these efforts. Needed disciplines include behavioral sciences; organizational behavior; intervention effectiveness research; public health education; dissemination, implementation, and diffusion research; social marketing; and media advocacy.

THE CHANGING NATURE OF WORK

Recognizing the changing demographics of the U.S. workforce, NIOSH has included reducing injuries among high-risk and vulnerable populations among its strategic goals for the future. It is likewise important to address the changing nature of work itself and its interplay with the changing workforce. Work in the United States is changing in significant ways that can be expected to alter the current pattern of and risk factors for work-related traumatic injury. The industrial sectors in which U.S. workers are employed are changing. Most notably, the United States continues to shift generally from a manufacturing to a service and knowledge economy. For example, the greatest growth is projected to occur in home health care, an industry which relies heavily on immigrant and minority labor and in which the work setting is geographically dispersed. Another example is where the "craft" of residential construction is changing to a manufactured or prefabricated industry. In addition, there is likely to be a continued shift in work organization and employment practices including corporate restructuring and downsizing, shifts to leaner, more flexible production methods, and increased reliance on part-time, temporary, and contingent labor. These trends may influence work hours, job demands, benefits, and job security that may in turn adversely impact injury risks and may disproportionately affect vulnerable worker populations. Recent emphasis on the development of new sustainable technologies and green building practices offers important new opportunities for prevention through design in which the health of working people as well as the environment would be taken into account in the design stage of new products and projects.

NIOSH has a cross-sector program on work organization and stress-related disorders which was also one of the 21 priority areas for research under the initial National Occupational Research Agenda (NORA). This program clearly recognizes the potential impact of the changing organization of work not only on worker health but also worker safety and has developed a research agenda to identify and address these potential risks (NIOSH, 2002). More recently NIOSH has established a program on prevention through design, which is broadly defined as addressing occupational safety and health needs in the design process to prevent or minimize work-related hazards and risks. The committee underscores the importance of TI Research Program collaboration with these other NIOSH program areas as well as the NIOSH Program on Occupational Health Disparities.

9. **Research prevention strategies for traumatic injuries in a changing workplace.** The TI Research Program should consider research on the safety impacts of changes in the nature of work as well as intervention research targeting organization polices and practices and including prevention through design approaches.

SUMMARY

Occupational traumatic injuries place a substantial burden on the country. The TI Research Program has demonstrated success in addressing them. Every program can improve, however, and NIOSH specifically requested that the ERCs provide suggestions for program improvement. The committee has offered nine recommendations in the areas of strategic planning, coordination and collaboration, workforce development, transfer, and the changing nature of work. The TI Research Program is aware of the need for these activities and is working to address all of them. The committee hopes that these recommendations are useful in supporting and encouraging its efforts.

With a focus on program improvement as outlined in this chapter, the TI Research Program can continue to serve as a leader in the field by identifying its niche in research, collaborating with partners, and sponsoring important high-quality research that contributes to reducing the morbidity and mortality associated with injury in the workplace.

REFERENCES

NIOSH (National Institute of Occupational Safety and Health). 2002. *The Changing Organization of Work and the Safety and Health of Working People: Knowledge of Gaps and Research Directions.* Washington, DC: HHS.

NIOSH. 2008. *Occupational health disparities.* http://www.cdc.gov/niosh/programs/ohd/risks.html (accessed July 21, 2008).

A

Framework for the Review of Research Programs of the National Institute for Occupational Safety and Health*

This is the second version of a document prepared by the National Academies Committee for the Review of NIOSH Research Programs[1] also referred to as the Framework Committee. This document is not a formal report of the National Academies—rather, it is a framework proposed for use by multiple National Academies evaluation committees to review up to 15 National Institute for Occupational Safety and Health (NIOSH) research programs. It is a working document subject to modification by the Framework Committee on the basis of responses received from evaluation-committee members, NIOSH, stakeholders, and the general public during the course of the assessments.

---

*Version of 8/10/07.

[1]Members of the committee at the time this version was produced were David Wegman, *Chair* (University of Massachusetts Lowell School of Health and Environment), William Bunn III (International Truck and Engine Corporation), Carlos Camargo (Harvard Medical School), Susan Cozzens (Georgia Institute of Technology), Letitia Davis (Massachusetts Department of Public Health), James Dearing (Kaiser Permanente-Colorado), Fred Mettler, Jr. (University of New Mexico School of Medicine), Franklin Mirer (Hunter College School of Health Sciences), Jacqueline Nowell (United Food and Commercial Workers International Union), Raja Ramani (Pennsylvania State University), Jorma Rantanen (International Commission on Occupational Health), Rosemary Sokas (University of Illinois at Chicago School of Public Health), Richard Tucker (Tucker and Tucker Consultants, Inc., and University of Texas at Austin), and James Zuiches (North Carolina State University). Sammantha Magsino (National Academies staff) was the study director. Joseph Wholey (University of Southern California), former committee member, contributed to the first version of this document. Part V includes brief biographies of current committee members.

This version reflects several significant changes to the original framework document (version 12/19/05) that was used to guide the work of the first four evaluation committees (Hearing Loss; Mining; Agriculture, Forestry, and Fishing; and Respiratory Disease). Changes were made in response to feedback from members and staff of these committees, as well as other comments on the original framework, in order to make the document more useful to evaluation committees as they carry out their work. In particular, the following changes were made to the framework document during the revision process:

- The wording of some of the relevance and impact scores were edited to make the wording more precise and to reduce situations where the original scores were non-unique or overlapping (revised scoring criteria are given in Boxes 2 and 3).
- A new table was added to provide explicit guidance to evaluation committees on how to weigh differences in the observed levels of "research priority" and "engagement in appropriate transfer activities" in arriving at a single integer score for relevance (see Table 6).
- The guidance on scoring was clarified to make more explicit that all scores are to be given as integers.
- The NIOSH logic model was updated (see Figure 1).
- The table on evaluation committee information needs (Table 2) was reorganized to be more consistent with the NIOSH logic model, and additional information needs identified by the first set of evaluation committees were added.
- A worksheet to assist with the development of scores has been deleted and key components of the worksheet have been incorporated into appropriate sections throughout the document.
- The organization of the document was modified to more closely follow the revised statement of task and to improve readability.
- A number of sections of text originally presented in outline form were modified in tables or boxes to make the information more accessible.

This second version of the framework document remains a working document subject to further modification by the Framework Committee on the basis of input received from evaluation committee members, NIOSH, stakeholders, and the general public during the course of the assessments.

CONTENTS

ABBREVIATIONS AND ACRONYMS

ABLES	Adult Blood Lead Epidemiology and Surveillance
AOEC	Association of Occupational and Environmental Clinics
BLS	Bureau of Labor Statistics
CDC	Centers for Disease Control and Prevention
CSTE	Council of State and Territorial Epidemiologists
DOD	U.S. Department of Defense
EC	Evaluation Committee
EPA	Environmental Protection Agency
FACE	Fatality Assessment Control and Evaluation
FC	Framework Committee
HHE	Health Hazard Evaluation
MSHA	Mine Safety and Health Administration
NIH	National Institutes of Health
NIOSH	National Institute for Occupational Safety and Health
NORA	National Occupational Research Agenda
NORA1	National Occupational Research Agenda 1996-2005
NORA2	National Occupational Research Agenda 2005-forward
OSH Review Commission	Occupational Safety and Health Review Commission
OSHA	Occupational Safety and Health Administration
OSHAct	Occupational Safety and Health Act of 1970
PART	Performance Assessment Rating Tool
PEL	permissible exposure limit
RFA	request for applications
SENSOR	Sentinel Event Notification System of Occupational Risks

TMT tools, methods, or technologies

USDA U.S. Department of Agriculture

I. INTRODUCTION

In September 2004, the National Institute for Occupational Safety and Health (NIOSH) contracted with the National Academies to conduct a review of NIOSH research programs. The goal of this multiphase effort is to assist NIOSH in increasing the impact of its research efforts that are aimed at reducing workplace illnesses and injuries and improving occupational safety and health. The National Academies assigned the task to the Division on Earth and Life Studies and the Institute of Medicine.

The National Academies appointed a committee of 14 members, including persons with expertise in occupational medicine and health, industrial health and safety, industrial hygiene, epidemiology, civil and mining engineering, sociology, program evaluation, communication, and toxicology; representatives of industry and of the workforce; and a scientist experienced in international occupational-health issues. The Committee on the Review of NIOSH Research Programs, referred to as the Framework Committee (FC), prepared the first version of this document during meetings held on May 5-6, July 7-8, and August 15-16, 2005. This second version was finalized after the Framework Committee's May 30-31, 2007, meeting, based on feedback received on the framework from the first two independent evaluation committees, NIOSH leadership, and National Academies' staff, as well as discussions during an earlier FC meeting in April 2006.

This document is not a report of the National Academies; rather, it presents the evaluation framework developed by the FC to guide and provide common structure for the reviews of as many as 15 NIOSH programs during a 5-year period by independent evaluation committees (ECs) appointed by various divisions and boards of the National Academies. It is a working document to be shared with NIOSH and the public. This version has been modified by the FC on the basis of responses from the ECs, NIOSH, NIOSH stakeholders, and the public; and it may be modified again. It is incumbent on the ECs to consult with the FC if portions of the evaluation framework presented here are inappropriate for specific programs under review.

I.A. Overview of Charge to Evaluation Committees

At the first meeting of the FC, Lewis Wade, NIOSH senior science adviser, emphasized that a review of a NIOSH program should focus on the program's relevance to and impact on health and safety in the workplace. In developing a framework, the FC considered the following elements of the charge to the ECs:

1. Assessment of the program's contribution, through occupational safety and health research, to reductions in workplace hazardous exposures, illnesses, or injuries through

   a. an assessment of the relevance of the program's activities to the improvement of occupational safety and health, and
   b. an evaluation of the impact that the program's research has had in reducing work-related hazardous exposures, illnesses, and injuries.

   The evaluation committee will rate the performance of the program for its relevance and impact using an integer score of 1-5. Impact may be assessed directly (for example, on the basis of reductions in illnesses or injuries) or, as necessary, by using intermediate outcomes to estimate impact. Qualitative narrative evaluations should be included to explain the numerical ratings.
2. Assessment of the program's effectiveness in targeting new research areas and identifying emerging issues in occupational safety and health most relevant to future improvements in workplace protection. The committee will provide a qualitative narrative assessment of the program's efforts and suggestions about emerging issues that the program should be prepared to address.

I.B. Evaluation Committees

Individual ECs will be formed in accordance with the rules of the National Academies for the formation of balanced committees. Each EC will comprise persons with expertise appropriate for the specific NIOSH research program under review and may include representatives of stakeholder groups (such as labor unions and industry), experts in technology and knowledge transfer, and program evaluation. The EC will gather appropriate information from the sponsor (the NIOSH research program under review), stakeholders affected directly by NIOSH program research, and relevant independent parties. Each EC will consist of about 10 members, will meet about three times, and will prepare a report. The National Academies will deliver the report to NIOSH within 9 months of the first meeting of the EC. EC reports are subject to the National Academies report-review process.

I.C. NIOSH Strategic Goals and Operational Plan

As a prelude to understanding the NIOSH strategic goals and operational plan, NIOSH research efforts should be understood in the context of the Occupational

Safety and Health Act (OSHAct), under which it was created. The OSHAct identifies workplace safety and health as having high national priority and gives employers the responsibility for controlling hazards and preventing workplace injury and illness. The act creates an organizational framework for doing that, assigning complementary roles and responsibilities to employers and employees, the Occupational Safety and Health Administration (OSHA), the states, the Occupational Safety and Health (OSH) Review Commission, and NIOSH. The act recognizes NIOSH's role and responsibilities to be supportive and indirect. NIOSH research, training programs, criteria, and recommendations are intended to be used to inform and assist those more directly responsible for hazard control (OSHAct Sections 2b, 20, and 22).

Section 2b of the OSHAct describes 13 interdependent means of accomplishing the national goal, one of which is "by providing for research . . . and by developing innovative methods . . . for dealing with occupational safety and health problems." Sections 20 and 22 give the responsibility for that research to NIOSH. NIOSH is also given related responsibilities, including the development of criteria to guide prevention of work-related injury or illness; development of regulations for reporting on employee exposures to harmful agents; establishment of medical examinations, programs, or tests to determine illness incidence and susceptibility; publication of a list of all known toxic substances; assessment of potential toxic effects or risks associated with workplace exposure in specific settings; and conduct of education programs for relevant professionals to carry out the OSHAct purposes. NIOSH is also responsible for assisting the secretary of labor regarding education programs for employees and employers in hazard recognition and control.

The NIOSH mission is "to provide national and world leadership to prevent work-related illness, injury, disability, and death by gathering information, conducting scientific research, and translating the knowledge gained into products and services." To fulfill its mission, NIOSH has established the following strategic goals:[2]

- **Goal 1: Conduct research to reduce work-related illnesses and injuries.**
  - Track work-related hazards, exposures, illnesses, and injuries for prevention.
  - Generate new knowledge through intramural and extramural research programs.

[2]See http://www.cdc.gov/niosh/docs/strategic/.

- Develop innovative solutions for difficult-to-solve problems in high-risk industrial sectors.

- **Goal 2: Promote safe and healthy workplaces through interventions, recommendations, and capacity building.**
  - Enhance the relevance and utility of recommendations and guidance.
  - Transfer research findings, technologies, and information into practice.
  - Build capacity to address traditional and emerging hazards.

- **Goal 3: Enhance global workplace safety and health through international collaborations.**
  - Take a leadership role in developing a global network of occupational health centers.
  - Investigate alternative approaches to workplace illness and injury reduction and provide technical assistance to put solutions in place.
  - Build global professional capacity to address workplace hazards through training, information sharing, and research experience.

In 1994, NIOSH embarked on a national partnership effort to identify research priorities to guide occupational health and safety research for the next decade. The National Occupational Research Agenda (NORA) identified 21 high-priority research subjects (see Table 1). The NORA was intended not only for NIOSH but for the entire occupational health community. In the second decade of the NORA, NIOSH is working with its partners to update the research agenda, using an approach based on industry sectors. NIOSH and its partners are working through sector research councils to establish sector-specific research goals and objectives. The emphasis is on moving research to practice in workplaces through sector-based partnerships.

Figure 1 is the NIOSH operational plan, presented as a logic model,[3] of the path from inputs to outcomes for each NIOSH research program. The FC adapted the model to develop its framework. NIOSH will provide similar logic models appropriate to each research program evaluated by an EC.

[3]Developed by NIOSH with the assistance of the RAND Corporation.

TABLE 1 NORA High-Priority Research by Category

Category	Priority Research Area
Disease and injury	Allergic and irritant dermatitis
	Asthma and chronic obstructive pulmonary disease
	Fertility and pregnancy abnormalities
	Hearing loss
	Infectious diseases
	Low-back disorders
	Musculoskeletal disorders of upper extremities
	Trauma
Work environment and workforce	Emerging technologies
	Indoor environment
	Mixed exposures
	Organization of work
	Special populations at risk
Research tools and approaches	Cancer research methods
	Control technology and personal protective equipment
	Exposure-assessment methods
	Health-services research
	Intervention-effectiveness research
	Risk-assessment methods
	Social and economic consequences of workplace illness and injury
	Surveillance research methods

I.D. Evaluation Committees' Information Needs

Each NIOSH program under review will provide information to the relevant EC, including that outlined in Table 2. The EC may request additional information of NIOSH as needed, and NIOSH should provide it as quickly as is practical. NIOSH should consider organizing the information listed in Table 2 by subprogram or program as appropriate and to the extent possible.

In addition to the information provided by NIOSH, the EC should independently collect additional information that it deems necessary for evaluation (for example, the perspectives of stakeholders, such as OSHA, MSHA, unions and workforces, and industry). In conducting the review, the EC should continually examine how individual projects or activities contribute to the impact and relevance of a program as a whole.

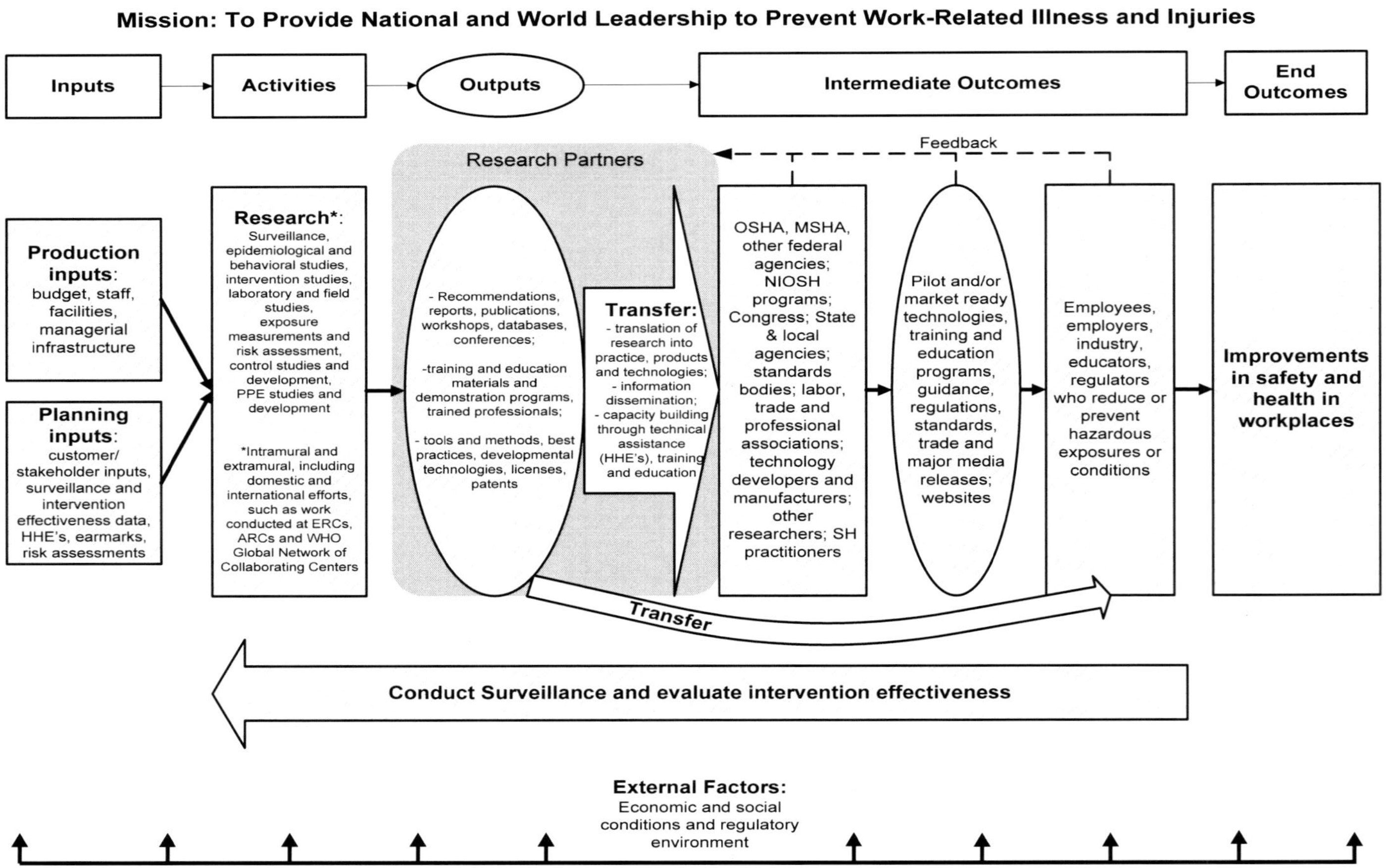

FIGURE 1 The NIOSH operational plan presented as a logic model.

TABLE 2 Evaluation Committee Information Needs

- **Program background and resources:**
  - Program history.
  - Major program challenges.
  - Program strategic goals and objectives, past (for period under review) and current.
  - Major subprograms (if appropriate).
  - Results of previous program reviews (for example, annual review by NIOSH leadership team or external scientific program reviews).
  - External factors affecting the program.

- **Interactions with stakeholders and with other NIOSH programs:**
  - The role of program research staff in NIOSH policy-setting, OSHA and MSHA standard-setting, voluntary standard-setting and other government policy functions.
  - Interactions and working relationships with other NIOSH programs.
  - Identification of other institutions and research programs with overlapping or similar portfolios and an explanation of the relationship between NIOSH activities and those of other institutions.
  - Key partnerships with employers, labor, other government organizations, academic institutions, nonprofit organizations, and international organizations.

- **Program inputs:**
  - Program resources (also called *production inputs*).
    - Funding by year for period under review.
    - Funding by objective or subprogram.
    - Program staffing, FTE's, and laboratory facilities, by subprogram (if indicated).
    - Percentage of program budget that is discretionary (beyond salaries).
    - Percentage of program budget that is earmarked.
    - Contributions from other agencies (in kind or funds).
  - Planning inputs.
    - Surveillance data, inputs from the Health Hazard Evaluation (HHE) or Fatality Assessment Control and Evaluation (FACE) program, or intramural and extramural research findings that influenced program goals and objectives.
    - Planning inputs from stakeholders, for example, advisory groups, NORA teams, and professional, industry, and labor groups (specify if any input from groups representing small business or vulnerable populations).
    - Related OSHA, Mine Safety and Health Administration (MSHA) strategic plans, or other input.
    - Process for soliciting and approving intramural research ideas.
    - Process for soliciting and approving program-supported extramural research activities.

- **Program activities (more details provided in Table 3):**
  - Intramural.
    - Surveillance activities.
    - Research activities (projects).

continued

TABLE 2 Continued

- Transfer activities to encourage implementation of research results for improved occupational safety and health (for example, information dissemination, technical assistance, and technology and knowledge transfer).
- Key collaborations in intramural activities (for example, with other government agencies, academe, industry, and unions).

- Extramural funded by NIOSH.
    - Requests for applications (RFAs) developed by program.
    - Funded projects: grants, cooperative agreements, and contracts, such as
        - ◊ Surveillance activities.
        - ◊ Research activities.
        - ◊ Transfer activities.
        - ◊ Capacity-building activities.

- **Outputs (products of the research program—more details provided in Table 4):**
    - Intramural.
        - Peer-reviewed publications, agency reports, alerts, and recommendations.
        - Databases, Web sites, tools, and methods (including education and training materials).
        - Technologies developed and patents.
        - Sponsored conferences and workshops.
    - Extramural (to the extent practical).

- **Intermediate outcomes:**
    - Standards or guidelines issued by other agencies or organizations based in whole or in part on NIOSH research.
    - Adoption and use of control or personal protective technologies developed by NIOSH.
    - Evidence of industry, employer, or worker behavioral changes in response to research outputs.
    - Use of NIOSH products by workers, industry, occupational health and safety professionals, health care providers, and so on (including internationally).
    - NIOSH Web-site hits and document requests.
    - Unique staff or laboratory capabilities that serve as a national resource.
    - Other intermediate outcomes.

- **End outcomes:**
    - Data on program impact on rates and numbers of injuries and illnesses and exposures in the workplace (including trend data, if available).
    - Documentation of workplace risk reduction (quantitative, qualitative, or both).

- **Description of current processes for setting research priorities and identifying emerging issues in the workplace.**

I.E. Prior Evaluations

Several NIOSH programs have already been evaluated by internal and external bodies. The evaluations may have been part of an overall assessment of NIOSH, such as the 2005 Performance Assessment Rating Tool (PART) review,[4] or the evaluation of specific research program elements, such as any external scientific-program review. NIOSH should inform of, and the ECs should review, all prior evaluations of the program under review as an aid to understanding the evolution of the program and its elements. The EC evaluations, however, are independent of prior reviews and evaluations.

II. SUMMARY OF EVALUATION PROCESS

The ECs will assess the relevance and impact of NIOSH research programs. In conducting their evaluations, the ECs should ascertain whether NIOSH is doing the right things (relevance) and whether these things are improving health and safety in the workplace (impact).

II.A. The Evaluation Flowchart (Figure 2)

To address its charge, the FC simplified the logic model of Figure 1 into a flowchart (Figure 2) that breaks the NIOSH logic model into discrete, sequential program components to be assessed by the EC. Each component of Figure 2 is addressed in greater detail in the indicated section of this document. The FC understands that the activities of any research program will not be as linear as presented in either Figures 1 or 2. The major components to be evaluated are

- major program *challenges,*
- strategic *goals and objectives,*
- *inputs* (such as budget, staff, facilities, the institute's research management, the NIOSH Board of Scientific Counselors, the NORA process, and NORA work groups),
- *activities* (efforts by NIOSH staff, contractors, and grantees, such as hazard surveillance; surveillance for injury, illness, and biomarkers of effect; exposure-measurement research; safety-systems research; injury-prevention research; health-effects research; intervention

[4]The PART focuses on assessing program-level performance and is one of the measures of success of the budget and performance integration initiative of the president's management agenda (see CDC Occupational Safety and Health at http://www.whitehouse.gov/omb/budget/fy2006/pma/hhs.pdf).

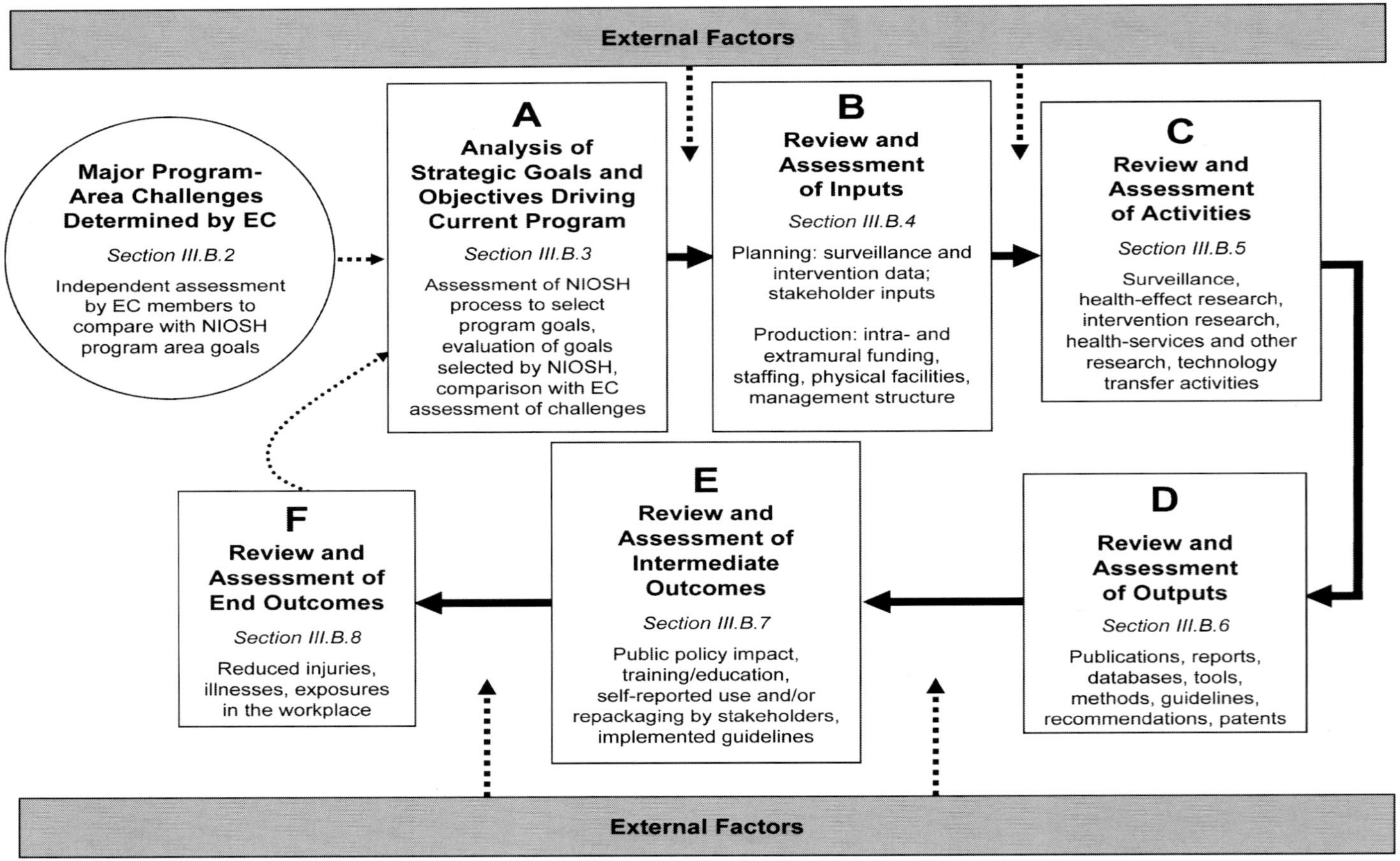

FIGURE 2 Flowchart for the evaluation of the NIOSH research program.

research; health-services research; and technology and knowledge transfer activities),
- *outputs* (NIOSH products, such as publications, reports, conferences, databases, tools, methods, guidelines, recommendations, education and training, and patents),
- *intermediate outcomes* (responses by NIOSH stakeholders to NIOSH products, such as public or private policy change, training and education in the form of workshop or seminar attendance, self-reported use or repackaging of NIOSH data by stakeholders, adoption of NIOSH-developed technologies, implemented guidelines, licenses, and reduction in workplace hazardous exposure), and
- *end outcomes* (such as reduction in work-related injuries or illnesses or hazardous exposures in the workplace).

The flowchart summarizes the FC's vision of how a program evaluation should occur. In evaluating each program or major subprogram, the EC must collect, analyze, and evaluate information on items described in each of the boxes of Figure 2, regardless of management structure (such as linear or matrix). The FC recognizes that the components of any program will not fit perfectly in any category in Figure 1 or 2. For example, training and development programs were appropriately defined as outputs by NIOSH in the logic model (Figure 1), but the FC finds more value in focusing on the responses to these outputs as intermediate outcomes (Figure 2, Box E) in the flowchart. The committee further recognizes that matrix organizations may have little control over the input portion of the logic model and that matrix program management may have fewer resources of its own on which to base its decisions. Following the suggested evaluation procedures, however, should ensure a desired level of consistency and comparability among all the ECs.

Drawing on the program logic model, the flowchart, and EC members' expertise, the ECs will delineate important inputs and external factors affecting the NIOSH research program's agenda and the consequences of NIOSH research activities. Examples of external factors are research activities of industry and other federal agencies and the political and regulatory environment. For purposes of this review, the results of inputs and external factors are the program research activities, outputs, and associated transfer activities that may result in intermediate outcomes and possibly end outcomes.

II.B. Steps in Program Evaluation

The FC concludes that useful evaluation requires a disciplined focus on a small number of questions or hypotheses typically related to program goals, performance criteria, and performance standards; a rigorous method of answering the questions or testing the hypotheses; and a credible procedure for developing qualitative and quantitative assessments. The evaluation process developed by the FC is summarized in Box 1 and described in detail in Section III of this document.

III. EVALUATION OF A NIOSH RESEARCH PROGRAM—THE PROCESS

III.A. Analysis of External Factors Relevant to the NIOSH Research Program

As depicted in the logic model (Figure 1), reduction in injury and illness (end outcomes) or in exposure (intermediate outcome) is affected by stakeholder activities (external factors). Actions of those in labor, industry, regulatory entities, and others beyond NIOSH's control are necessary for the implementation of NIOSH recommendations. Implementation of research findings may depend on existing or future policy considerations.

External factors may be considered as forces beyond the control of NIOSH that may affect the evolution of a program. External factors influence NIOSH's progress through all phases of the logic model and flowchart; from inputs to end outcomes (see Figures 1 and 2). Identification of external factors by an EC is essential because it provides the context for evaluation of the NIOSH program. External factors may be best assessed on the basis of the expert judgment of EC members who have knowledge of the field of research. Information regarding external factors should also be sought from NIOSH, OSHA, and MSHA staff and from other stakeholders. The EC, however, may choose additional approaches to assess external factors. NIOSH should identify and describe external factors early in the evaluation sequence (see Table 2). Factors external to NIOSH might have been responsible for achieving some outcomes or might have presented formidable obstacles. The EC must address both possibilities.

Some external factors may involve constraints on research activities related to target populations, methodologic issues, and resource availability. ECs might examine whether or not the following are true:

- Projects addressing a critical health need are technologically feasible. However, a workforce of appropriate size and with appropriate duration and distribution of exposure for measuring a health effect may not exist; for example, no population of workers has been exposed for 30 years to formaldehyde at the current OSHA permissible exposure limit (PEL), so the related cancer mortality cannot yet be directly assessed.
- Research is inhibited because NIOSH investigators are unable to access an adequate study population. Under current policy, NIOSH must

BOX 1
The Evaluation Process

1. Gather appropriate information from NIOSH and other sources (see Table 2).

2. Determine timeframe to be covered in the evaluation (see III.B.1.).

3. Identify major program area challenges and objectives (see III.B.2.).
   All NIOSH research programs are designed to be responsive to present or future workplace safety and health issues. Each research program should have its own objectives. Each EC will provide an independent assessment of the major workplace health and safety problems related to the program under review and determine whether they are consistent with the program's stated goals and objectives.

4. Identify subprograms and major projects in the research program.
   Each EC must determine how to disaggregate a program to achieve a manageable and meaningful evaluation of its components, and of the overall program. A program may need to be broken down into several recognizable subprograms or major projects if an effective evaluation is to be organized. It may be advantageous for an EC to disaggregate a program into subprograms that NIOSH identifies.

5. Evaluate the subprogram components sequentially (see III.B.2. through III.B.8.), using the flowchart (Figure 2) as a guide.
   This will involve a qualitative assessment of each component of the research program. ECs will use professional judgment to answer questions and follow the guidance provided by the FC.

6. Evaluate the research program's potential outcomes that are not yet appreciated (see III.B.9.).

7. Evaluate the important subprogram outcomes specifically for contributions to improvements in workplace safety and health.
   Guidance is provided with specific items for consideration (see III.B.10.).

8. Evaluate and score the overall program for *relevance* (see III.B.10.).
   Final program ratings will consist of an integer score and discussion of its rationale.

9. Evaluate and score the overall program for *impact* (see III.B.10.).
   Final program ratings will consist of an integer score and discussion of its rationale.

10. Identify success in targeting priority research and emerging issues (see III.C.).
    The EC should briefly discuss its assessment of the NIOSH program's process for determining priorities for research and emerging workplace issues. The ECs should also independently identify emerging workplace issues for which the NIOSH program under review should be prepared.

11. Prepare report by using the template provided in Section IV as a guide.

either obtain an invitation by management to study a workplace or seek a judicial order to provide authority to enter a worksite. (Cooperation under court order may well be insufficient for effective research.)

- Research is inhibited because the work environment, materials, and historical records cannot be accessed even with management and workforce cooperation.
- Adequate or established methods do not exist for assessing the environment.
- The NIOSH contribution to a particular field of research is reduced because other institutions are working in the same field.
- NIOSH resources are inadequate to tackle key questions.

Evaluation of the impact of NIOSH research outputs on worker health and safety may require consideration of external factors that might impede or aide implementation, measurement, and so on. ECs might consider whether or not the following are true:

- Regulatory end points are unachievable because of obstacles to regulation or because of differing priorities of the regulatory agencies. For example, there may be no implementation of recommendations for improved respiratory protection programs for health-care workers because of enforcement policies or lack of acceptance by the health-care institution administrators.
- A feasible control for a known risk factor or exposure is unimplemented because the costs of implementation are too high or because current economic incentives do not favor such actions.
- End outcomes are unobservable because baseline and continuing surveillance data are not available. For example, the current incidence of occupational noise-induced hearing loss is not known although surveillance for a substantial threshold shift is feasible. (NIOSH conducts surveillance of work-related illnesses, injuries, and hazards, but comprehensive surveillance is not possible with existing resources.)
- Reductions in adverse effects of chronic exposure cannot be measured. For example, 90% of identified work-related mortality is from diseases, such as cancer, that arise only after decades of latency after first exposure; therefore, effects of reducing exposure to a carcinogen cannot be observed in the timeframe of most interventions.
- A promulgated regulation requires a technology that was developed but not widely used.

- Reductions in fatal traumatic injuries occur because more-hazardous manufacturing jobs are replaced by less-hazardous knowledge-based jobs.

III.B. Evaluating NIOSH Research Programs by Using the Flowchart

The FC used the NIOSH logic model (Figure 1) to define the scope and stages of an EC evaluation. The evaluation of the elements in the flowchart (Figure 2) summarizes the FC's vision of how a program evaluation should proceed. FC members also identified numerous possible factors to consider in assessing the relevance of NIOSH research-program components, including the following:

- The severity or frequency of health and safety hazards addressed and the number of people at risk (magnitude) for these hazards.
- The extent to which NIOSH research programs identify and address gender-related issues and issues of vulnerable populations: Vulnerable populations are defined as groups of workers who have biologic, social, or economic characteristics that place them at increased risk for work-related conditions or on whom inadequate data have been collected. Vulnerable populations include disadvantaged minorities, disabled persons, low-wage workers, and non-English-speakers for whom language or other barriers present health or safety risks.
- The extent to which NIOSH research programs address the health and safety needs of small businesses.
- The "life stage" of problems being addressed: As the health effects are understood, efforts should shift to intervention research, from efficacy to intervention, and to intervention-effectiveness research. Gaps in the spectrum of prevention need to be addressed; for example, research on exposure assessment may be necessary before the next intervention steps can be taken.
- The structure, in addition to the content, of the research program: A relevant research program is more than a set of unrelated research projects; it is an integrated program involving interrelated surveillance, research, and transfer activities.
- Appropriate NIOSH consideration of stakeholder input.

The ECs may consider those and other important factors that bear on relevance as they progress through each stage of an evaluation.

The following subsections are intended to guide the EC through the evaluation process and flowchart in Figure 2. Each begins with a definition of the component being evaluated, provides questions for the EC to consider during the course of its evaluation, and provides some guidance regarding the assessment of the component. The FC admittedly provides little guidance regarding the evaluation of programs that are organized in a matrix structure or programs that have large extramural research components. Because of the uniqueness of each program, each EC must determine the most reasonable way to apply the criteria established in this document.

III.B.1. Identifying the Period for Evaluation

By studying materials presented by the NIOSH research program and other sources, the EC will become familiar with the history of the research program being evaluated and its major subprograms, goals, objectives, resources, and other pertinent information. Having that information, the EC should choose the period most appropriate for the evaluation. EC efforts should focus on the impact and relevance of the NIOSH program in the most recent appropriate period. As a starting point, the ECs might consider three general timeframes:

- 1970-1995, the period from the founding of NIOSH to the initiation of NORA (pre-NORA period)
- 1996-2005 (NORA 1 period)
- After 2005 (NORA 2 period)

Those timeframes are provided as general guidance; the period chosen for review will take into consideration suggestions from the NIOSH research program under review. It is recognized that many of the intermediate and end outcomes documented since 1996 are consequences of research outputs completed before 1996.

III.B.2. Identifying Major Challenges (Figure 2, Circle)

Early in the assessment process, the EC itself should identify the major workplace health and safety challenges for the research program under review. In arriving at a list of challenges, the EC should rely on surveillance findings, including those of NIOSH investigations of sentinel events (through health-hazard or fatality-assessment programs), external advisory inputs, and its own expert judgment. The EC will then be able to compare its own assessment of workplace challenges with the NIOSH program goals and objectives. The congruence between the two will be useful during the assessment of relevance.

III.B.3. Analysis of Research-Program Strategic Goals and Objectives (Figure 2, Box A)

The research program goals and objectives should be evaluated with a focus on how each program goal is related to NIOSH's agency wide strategic goals and to the program challenges identified in the step above (Section III.B.2.). The importance or relevance of an issue may differ from the influence of NIOSH-funded research in addressing it. The EC should recognize that NIOSH research priorities may be circumstantial (for example, congressionally funded) rather than based on NIOSH's assessment of the state of knowledge.

Questions to Guide the Evaluation Committee

1. Are the strategic goals and objectives of the program well defined and clearly described?
2. How well were program goals and objectives aligned with NORA 1 priorities during the last decade?
3. How are current program strategic goals and objectives related to current NIOSH strategy, including NORA 2?
4. Are the research program goals, objectives, and strategies relevant to the major challenges for the research program and likely to address emerging problems in the research program (as determined by the EC while addressing Section III.B.2.)?
   a. Did past program goals and objectives (research and dissemination and transfer activities) focus on the most relevant problems and anticipate the emerging problems in the research program?
   b. Do the current program goals and objectives target the most relevant problems?

Assessment

The EC should provide a qualitative assessment that discusses the relevance of the program's goals, objectives, and strategies in relation to its major challenges.

III.B.4. Review of Inputs (Figure 2, Box B)

Planning inputs include input from stakeholders, surveillance and intervention data, and risk assessments. Production inputs include intramural and extramural funding, staffing, management structure, and physical facilities.

The EC should examine existing intramural and extramural resources and, potentially, prior surveys or case studies that might have been developed specifically to assess progress in reducing workplace illnesses and injuries and to provide information relevant to the targeting of research to future needs. The NIOSH research program should provide the EC all relevant planning and production inputs (see below and Table 2 for examples).

Planning Inputs

Planning inputs can be qualitative or quantitative. Sources of qualitative inputs include the following:

- Federal advisory committees (such as the Board of Scientific Counselors, the Mine Safety and Health Research Advisory Committee, and the National Advisory Committee on Occupational Safety and Health)
- NORA research partners, initial NORA stakeholder meetings, later NORA team efforts (especially strategic research plans), and the NORA Liaison Committee and federal liaison committee recommendations
- Industry, labor, academe, professional associations, industry associations, and the Council of State and Territorial Epidemiologists (CSTE)
- OSHA and MSHA strategic plans and other federal research agendas

Attention should be given to how comprehensive the inputs have been and to what extent gaps in input have been identified and considered by NIOSH.

Sources of quantitative inputs include the following:

- Intramural surveillance information, such as descriptive data on exposures and outcomes (appropriate data may be available from a number of NIOSH divisions and laboratories)
- HHEs
- Reports from the FACE program
- Extramural health-outcome and exposure-assessment data from OSHA, MSHA (both safety and health inspection data), the Bureau of Labor Statistics, the U.S. Department of Defense (DOD), and the U.S. Department of Agriculture (USDA) (fatality, injury, and illness surveillance data); state government partners, including NIOSH-funded state surveillance programs, such as Sentinel Event Notification System of Occupational Risks (SENSOR), Adult Blood Lead Epidemiology and

Surveillance (ABLES), and state-based FACE; and nongovernment organizations, such as the National Safety Council, the Association of Occupational and Environmental Clinics (AOEC), the American Society of Safety Engineers, and the American College of Occupational and Environmental Medicine
- Appropriate data from investigator-initiated extramural research funded by NIOSH

Production Inputs

For the research program under review, NIOSH should identify portions of the NIOSH intramural budget, staff, facilities, and management that play major roles in the research program. Production inputs should be described primarily in terms of intramural research projects, relevant extramural projects (particularly cooperative agreements and contracts), HHEs, and related staff. Consideration should also be given to leveraged funds provided by such partners as the National Institutes of Health (NIH) and the Environmental Protection Agency (EPA) for joint requests for applications or program announcements; and to OSHA, MSHA, and U.S. Department of Defense (DOD) contracts with NIOSH.

Assessment of inputs should include EC consideration of the degree to which allocation of funding and personnel was commensurate with the resources needed to conduct the research and the extent to which funding for the relevant intramural research activity has been limited by lack of discretionary spending beyond salaries (travel, supplies, external laboratory services, and so on). Thus, assessments should consider the adequacy of the qualitative and quantitative planning and production inputs, given the tasks at hand.

Questions to Guide the Evaluation Committee

1. Do planning, production, and other input data promote program goals?
2. How well are major planning, production, and other program inputs used to support the major activities?
3. Is input obtained from stakeholders, including input representing vulnerable working populations and small businesses?
4. Are production inputs (intramural and extramural funding, staffing, management, and physical infrastructure resources) consistent with program goals and objectives?

Assessment

The EC should provide a qualitative assessment that discusses the quality, adequacy, and use of inputs.

III.B.5. Review of Activities (Figure 2, Box C)

Activities are defined as the efforts and work of a program's staff, grantees, and contractors. For present purposes, activities of the NIOSH program under review are divided into research and transfer activities. Table 3 is intended to guide the EC and NIOSH as to the type and organization of information required to evaluate program activities. The table may be incomplete, and some types of research activity may not be applicable to a given NIOSH program. Research activities include safety research, health-outcomes research, safety-design research, and safety-systems research. Transfer activities include information dissemination, training, technical assistance, and education designed to translate research outputs into content and formats that are designed for application in the workplace. Depending on the scope of the program under review, activities may also be grouped by research-program objectives or subprograms.

Conventional occupational safety and health research focuses appropriately on injury, illness, or death; on biomarkers of exposure; and on health effects of new technology, personal protective equipment, and regulations. A focus on surveillance research may be needed when available data inputs are inadequate. A focus on socioeconomic and policy research and on diffusion research is also needed to effect change, because not all relevant intermediate outcomes occur in the workplace. NIOSH may be able to affect important outcomes farther out on the causal chain so as to influence health and safety in the workplace. Other research that might prove important in addressing NIOSH's mission includes the following:

- Surveillance research to assess the degree of significant or systematic underreporting of relevant injuries, illnesses, and biomarkers
- Socioeconomic research on cost-shifting between worker compensation and private insurance
- Research on methods to build health and safety capacity in community health centers that serve low-income or minority-group workers and to improve recognition and treatment of work-related conditions
- Transfer research to change health and safety knowledge of adolescents while they are in high school to improve the likelihood of reduced injuries as they enter the workforce

TABLE 3 Examples of NIOSH Program Research and Transfer Activities

Surveillance (including hazard and injury, illness, and biomarkers of exposure or effect health surveillance and evaluation of surveillance systems)
Health-effects research (illnesses, injuries, and biomarkers):
Epidemiology
Toxicology
Physical and safety risk factors (laboratory-based)
Development of clinical-screening methods and tools
Exposure-assessment research:
Chemical hazards
Physical hazards
Biologic hazards
Ergonomic hazards
Safety (traumatic injury) hazards
Safer-design and safety-systems research
Intervention research:
Control technologies
Engineering controls and alternatives
Administrative controls
Personal protective equipment
Work organization
Community participation
Policy (such as alternative approaches to targeting inspections)
Design for safety
Emergency preparedness and disaster response
Diffusion and dissemination research:
Training effectiveness
Information-dissemination effectiveness
Diffusion of technology
Health-services and other research:
Access to occupational health care
Infrastructure—delivery of occupational-health services, including international health and safety
Socioeconomic consequences of work-related injuries and illnesses
Worker compensation
Technology-transfer and other transfer activities:
Information dissemination
Training programs
Technical assistance

- Community-based participatory research on differences between recently arrived immigrants and U.S.-born workers regarding perceptions of acceptable health and safety risks so that programs can be targeted to meet the workforce training needs of immigrant workers

Transfer activities should be reviewed to determine whether the NIOSH program appropriately targets its outputs in a manner that will have the greatest impact. Ideally, information dissemination should be proactive, and strategic dissemination should be informed by research on the diffusion of new technologies, processes, and practices. Highly relevant information and technology transfer should include plans for appropriate transfer to all appropriate worker populations, including those considered vulnerable. Training should be incorporated into the strategic goals of all research fields where appropriate.

The EC should review project-level research and transfer activities (including surveillance activities) that have been completed, are in progress, or planned by the program under review. The program under review should provide a list of activities and specify whether they are intramural or extramural. For each extramural project, the key organizations and principal investigators' names should be requested, as should whether the project was in response to a request for proposal or a request for application. For each intramural project, the EC should ask NIOSH to provide a list of key collaborators (from another government agency, academe, industry, or unions).

The EC should evaluate each of the research activities outlined in Table 3 if it forms an important element of the program research. In the case of a sector-based research program (for example, mining or construction) in which health-effects research is not being reviewed, the EC should determine what research inputs influence the program's strategic goals and objective, and then assess the value of the inputs.

Questions to Guide the Evaluation Committee in Assessing Research Activities

1. What are the major subprograms or groupings of activities within the program?
2. Are activities consistent with program goals and objectives?
3. Are research activities relevant to the major challenges of the research program?
    a. Do they address the most serious outcomes?
    b. Do they address the most common outcomes?
    c. Do they address the needs of both sexes, vulnerable working populations, and small businesses?

4. Are research activities appropriately responsive to the input of stakeholders?
5. To what extent are partners involved in the research activities?
6. Are partners involved early in the research process so that they could participate in determining research objectives and research design?
7. Were original resource allocations appropriate for the research activities, and do they remain appropriate?
8. To what extent does peer reviews (internal, external, and midcourse) affect the activities?
9. Is there adequate monitoring of quality-assurance procedures to ensure credible research data, analyses, and conclusions?

Questions to Guide the Evaluation Committee in Assessing Transfer Activities

1. Is there a coherent planned program of transfer activities?
2. Are the program's information dissemination, training, education, technical assistance, or publications successful in reaching the workplace or relevant stakeholders in other settings? How widespread is the response?
3. To what degree have stakeholders responded to NIOSH information and training products?
4. Is there evidence that the formats for information products were selected in response to stakeholder preferences?
5. To what extent do program personnel rely on assessment of stakeholder needs and reactions to prototype information and training projects (formative evaluation techniques)?
6. To what extent does the program build research and education capacity internally and among stakeholders?

Assessment

For this part of the assessment, the EC will provide a qualitative assessment that discusses relevance. This assessment should include consideration of the external factors identified in Section III.A. that constrain choices of research projects and the relevance and effectiveness of transfer activities. The EC should consider the appropriateness of resource allocations. A highly relevant program would address high-priority needs, produce high-quality results, be appropriately collaborative, be of value to stakeholders, and be substantially engaged in transfer activities. A program may be less relevant to the extent that those key elements are not up to the mark or are missing. The discussion should cover those aspects in sufficient detail to arrive at a qualitative assessment of the activities. Assessment of the transfer activities must include considerations of program planning, coherence, and impact. The EC might also consider the incorporation of international research results into NIOSH knowledge-transfer activities for industry sectors in the United States.

III.B.6. Review of Outputs (Figure 2, Box D)

An output is a direct product of a NIOSH research program. Outputs may be designed for researchers, practitioners, intermediaries, and end-users, such as consumers. Outputs can be in the form of publications in peer-reviewed journals, recommendations, reports, Web-site content, workshops and presentations, databases, educational materials, scales and methods, new technologies, patents, technical assistance, and so on. Outputs of the research program's extramurally funded activities should also be considered. Table 4 lists examples of major outputs to be considered by the EC. The NIOSH research program should make every effort to include all pertinent data of the types listed in the table.

Outputs may be tailored to the intended audience to communicate information most effectively and increase the likelihood of comprehension, knowledge, attitude formation, and behavioral intent. The extent of use of formative evaluation data (data gathered before communication for the purpose of improving the likelihood of the intended effects) and the extent of intended user feedback in the design of the output can be considered indicators of appropriate quality assessment.

Some activities such as collaborations can also legitimately be conceptualized as outputs, because the collaboration itself is a result of NIOSH efforts. Cooperation, coordination, more intensive collaboration, and eventual formal partnering can be considered important outputs leading to desirable intermediate outcomes. Technology and knowledge transfer is greatly facilitated through such relationships. The extent of collaboration with other organizations in the determination of research agendas, the conduct of research, the dissemination of research results, and interorganization involvement in the production of outputs can all be measures of output quality and quantity. The EC may consider coauthorship while trying to determine the importance of NIOSH research to the broader research community.

The NIOSH program should provide information on all relevant outputs of the program under review produced during the chosen period.

Questions to Guide the Evaluation Committee

1. What are the major outputs of the research program?
2. Are output levels consistent with resources allocated (were resources allocated and used efficiently to produce outputs)?

TABLE 4 Examples of Research-Program Outputs to Be Considered

Peer-reviewed publications by NIOSH staff:
- Number of original research articles by NIOSH staff
- Number of review articles by NIOSH staff (including best-practices articles)
- Complete citation for each publication
- Complete copies of the "top five" articles
- Collaboration with other public- or private-sector researchers
- Publications in the field of interest with other support by investigators also funded by NIOSH (for example, ergonomic studies with other support by an investigator funded by NIOSH to do ergonomics work, in which case NIOSH should get some credit for seeding interest or drawing people into the field)

Peer-reviewed publications by external researchers funded by NIOSH:
- Number of NIOSH-funded original research articles by external researchers
- Number of NIOSH-funded review articles by external researchers (including best-practices articles)
- Complete citation for each written report
- Complete copies of the "top five" articles
- Collaboration with other government or academic researchers

NIOSH reports in the research program:
- Number of written reports
- Complete citation for each written report
- Complete copies of the "top five" reports

Sponsored conferences and workshops:
- Number of sponsored conferences
- Number of sponsored workshops
- Description of conferences and workshops (title, date, sponsors, target audience, number of participants, and resulting products)
- NIOSH's assessment of value or impact

Databases:
- Number of major databases created by NIOSH staff
- Number of major databases created by external researchers funded by NIOSH grants
- Description of databases:
  - Title, objective (in one to four sentences), and start and stop dates
  - Partial vs. complete sponsorship (if partial, who were cosponsors?)
  - Study or surveillance-system design, study population, and sample size
  - Primary "products" of the database (such as number of peer-reviewed articles and reports)
- Complete copies of the "top two" publications or findings, to date, from each database

continued

TABLE 4 Continued

Recommendations:
- Number of major recommendations
- Description of recommendations:
  - Complete citation (article, report, or conference where recommendation was made)
  - Summary in one to four sentences
  - Percentage of target audience that has adopted recommendation 1, 5, and 10 years later
  - Up to three examples of implementation in the field
- Identification of "top five" recommendations to date

Tools, methods, and technologies (TMT):
- Number of major TMT (includes training and education materials)
- Descriptions of TMT
  - Title and objective of TMT (in one to four sentences)
  - Complete citation (if applicable)
  - Percentage of target audience that has used TMT 1, 5, and 10 years later
  - Up to three examples of implementation in the field
- Identification of "top five" TMT to date

Patents:
- Total number of patents
- For each:
  - Title and objective (in one to four sentences)
  - Complete citation
  - Percentage of target audience that has used product 1, 5, and 10 years later
  - Up to three examples of implementation in the field
- Identification of "top five" patents to date

Miscellaneous:
- Any other important program outputs

3. Does the research program produce outputs that address high-priority areas?
4. To what extent does the program generate important new knowledge or technology?
5. Are there widely cited peer-reviewed publications considered to report "breakthrough" results?
6. What, if any, internal or external capacity-building outputs are documented?
7. Are outputs relevant to both sexes, vulnerable populations, and do they address health disparities?
8. Are outputs relevant to health and safety problems of small businesses?

9. Are products user-friendly with respect to readability, simplicity, and design?
10. To what extent does the program help to build the internal or extramural institutional knowledge base?
11. Does the research produce effective cross-agency, cross-institute, or internal-external collaborations?
12. To what extent does the program build research and education capacity (internal or external)?

Assessment

The EC should provide a qualitative assessment discussing relevance and utility. The outputs of a highly ranked program will address needs in high-priority areas, contain new knowledge or technology that is effectively communicated, contribute to capacity-building inside and outside NIOSH, and be relevant to the pertinent populations. The discussion should cover those aspects in sufficient detail to support the qualitative assessment of the outputs.

III.B.7. Review of Intermediate Outcomes (Figure 2, Box E)

Intermediate outcomes are important indicators of stakeholder response to NIOSH outputs. They reflect the impact of program activities and may lead to the desired end outcome of improved workplace safety and health. Intermediate outcomes include the production by those outside of NIOSH of guidelines or regulations based wholly or partly on NIOSH research (products adopted as national or international public policy or as policy or guidelines by private organizations or industry); contributions to training and education programs sponsored by other organizations; use of publications or other materials by workers, industry, and occupational safety and health professionals in the field; and citations of NIOSH research by industrial and academic scientists.

Intermediate outcomes allow inference that a program's outputs are associated with observed changes in the workplace. Thus, an intermediate outcome reflects an assessment of worth by NIOSH stakeholders (such as managers in industrial firms) about NIOSH research or its products (for example, NIOSH training workshops). Intermediate outcomes that are difficult to monitor but may be valid indicators of relevance or utility include self-report measures by users of NIOSH outputs. Such indicators include the extent to which key intermediaries find value in NIOSH products or databases for the repackaging of health and safety information, the extent to which NIOSH recommendations are in place and attended to in workplaces, and employee or employer knowledge of and adherence to NIOSH-recommended practices.

Questions to Guide the Evaluation Committee

1. Do program outputs result in or contribute to stakeholder training or education activities used in the workplace or in school or apprentice programs? If so, how?
2. Do program activities and outputs result in regulations, public policy, or voluntary standards or guidelines that are transferred to or created by the workplace?
3. Has the program resulted in changes in employer or worker practices associated with the reduction of risk (for example, in the adoption of new feasible control or personal protective technologies or administrative control concepts)?
4. Does the program contribute to changes in health-care practices to improve recognition and management of occupational health conditions?
5. Does the program result in research partnerships with stakeholders that lead to changes in the workplace?
6. To what extent do the program's stakeholders find value in NIOSH products (as shown by document requests, Web-site hits, conference attendance, and so on)?
7. Does the program or a subprogram provide unique staff or laboratory capability that is a necessary national resource? If so, is it adequate, or does it need to be enhanced or reduced?
8. Has the program resulted in interventions that protect both sexes, vulnerable workers, or address the needs of small businesses?
9. To what extent did the program contribute to increased capacity at worksites to identify or respond to safety and health threats?

Assessment

Only a qualitative assessment of product development, usefulness, and impact is required at this point in the EC report. Some thought should be given to the relative value of intermediate outcomes, and the FC recommends applying the well-accepted hierarchy-of-controls model. The discussion could include comments on how widely products have been used or programs implemented. The qualitative discussion should be specific as to the various products developed by the program and the extent of their use by specific entities (industry, labor, government, and so on) for specific purposes. Whether the products have resulted in changes in the workplace or in the reduction of risk should be discussed. The recognition accorded to the program or the facilities by its peers (such as recognition as a "center of excellence" by national and international communities) should be considered in the assessment. To be highly ranked, a program should have high performance in most of the relevant questions in this section. An aspect of the evaluation can be whether the same changes in stakeholder activities and behaviors would probably have occurred without NIOSH efforts.

III.B.8. Review of End Outcomes (Figure 2, Box F)

It is necessary for the EC to assess, to the greatest extent possible, NIOSH's contribution to end outcomes—improvements in workplace health and safety (impact). For purposes of this evaluation, end outcomes are health-related changes that are a result of program activities, including decreases in injuries, illnesses, deaths and exposures or risk. Data on reductions in work-related injuries, illnesses, and hazardous exposures will be available for some programs, and in some cases they will be quantifiable. It is possible, however, to evaluate the impact of a NIOSH research program using either intermediate outcomes or end outcomes. If there is no direct evidence of improvements in health and safety, intermediate outcomes may be used as proxies for end outcomes in assessing impact as long as the EC qualifies its findings. The EC will describe the realized or potential benefits of the NIOSH program. Examples of realized intermediate outcomes are new regulations and widely accepted guidelines, work practices, and procedures, all of which may contribute measurably to enhancing health and safety in the workplace.

The FC recognizes that assessing the causal relationship between NIOSH research and specific occupational health and safety outcomes is a major challenge because NIOSH does not have direct responsibility or authority for implementing its research findings in the workplace. Furthermore, the benefits of NIOSH research program outputs can be realized, potential, or limited to the knowledge gained. Studies that conclude with negative results may nevertheless have incorporated excellent science and contribute to the knowledge base. The generation of important knowledge is a recognized form of outcome in the absence of measurable impacts.

The impact of an outcome depends on the existence of a "receptor" for research results, such as a regulatory agency, a professional organization, an employer, and an employee organization. The EC should consider questions related to the various stages that lead to outputs, such as these:

1. Did NIOSH research identify a gap in protection or a means of reducing risk?
2. Did NIOSH convey that information to potential users in a usable form?
3. Were NIOSH research results (for example, recommendations, technologies) applied?
4. Did the applied results lead to desired outcomes?

Quantitative data are preferable to qualitative, but qualitative analysis may be necessary. Sources of quantitative data include the following:

- Bureau of Labor Statistics (BLS) data on fatal occupational injuries (the Census of Fatal Occupational Injuries) and nonfatal occupational injuries and illnesses (the annual Survey of Occupational Injury and Illnesses)
- NIOSH intramural surveillance systems, such as the National Electronic Injury Surveillance System, the coal-worker x-ray surveillance program, and agricultural-worker surveys conducted by NIOSH in collaboration with USDA
- State-based surveillance systems, such as the NIOSH-funded ABLES, and the SENSOR programs (for asthma, pesticides, silicosis, noise-induced hearing loss, dermatitis, and burns)
- Selected state worker-compensation programs
- Exposure data collected in the OSHA Integrated Management Information System

The FC is unaware of mechanisms for surveillance of many occupationally related chronic illnesses, such as cancers that arise from long exposure to chemicals and other stressors. The incidence and prevalence of many such outcomes are best evaluated by investigator-initiated research. Research that leads to new, effective surveillance concepts or programs warrants special recognition.

The EC should recognize the strengths and weaknesses of outcome data sources. Quantitative accident, injury, illness, and employment data and databases are subject to error and bias and should be used by the EC only for drawing inferences after critical evaluation and examination of available corroborating data. For example, it is widely recognized that occupational illnesses are poorly documented in the BLS Survey of Occupational Injuries and Illnesses, which captures only incident cases among active workers. It is difficult for health practitioners to diagnose work-relatedness of most illnesses that may not be exclusively related to work; furthermore, few practitioners are adequately trained to make such an assessment. Many of those illnesses have long latencies and do not appear until years after people have left the employment in question. Surveillance programs may systematically undercount some categories of workers, such as contingent workers.

In addition to measures of illness and injury, measures of exposure to chemical and physical agents and to safety and ergonomic hazards can be useful. Exposure or probability of exposure can serve as an appropriate proxy for disease or injury when a well-described occupational exposure-health association exists. In such instances, a decrease in exposure can be accepted as evidence that the end outcome of reduced illness or injury is being achieved. That is necessary particularly when the latent period between exposure and disease outcome, as in the case of asbestos exposure and lung cancer, makes effective evaluation of the relevant end outcome infeasible.

As an example of how an exposure level can serve as a proxy, reduction in the number of sites that exceed an OSHA PEL or an American Conference of Governmental Industrial Hygienists threshold limit value is a quantitative measure of improvement of occupational health awareness and reduction of risk. In addition to exposure level, the number of people exposed and the distribution of exposure levels are important. Those data are available from multiple databases and studies of exposure. Apart from air monitoring, such measures of exposure as biohazard controls, reduction in requirements for use of personal protective equipment, and reduction in ergonomic risks are important.

Challenges posed by inadequate or inaccurate measurement systems should not drive programs out of difficult fields of study, and the EC will need to be aware of such a possibility. In particular, contingent and informal working arrangements that place workers at greatest risk are also those on which surveillance information is almost totally lacking, so novel methods for measuring impact may be required.

The commitment of industry, labor, and government to health and safety are critical external factors. Several measures of that commitment can be useful for the EC: monetary commitments, attitude, staffing, and surveys of relative importance. To the extent that resources allocated to safety and health are limiting factors, the EC should explicitly assess NIOSH performance in the context of constraints.

Questions to Guide the Evaluation Committee

1. What are the amounts and qualities of relevant end-outcomes data (such as injuries, illness, exposure, and productivity affected by health)?
2. What are the temporal trends in those data?
3. Is there objective evidence of improvement in occupational safety or health?

4. To what degree is the NIOSH program or subprogram responsible for improvement in occupational safety or health?
5. If there is no time trend in the data, how do findings compare with data from other comparable U.S. groups or the corresponding populations in other countries?
6. What is the evidence that external factors have affected outcomes or outcome measures?
7. Has the program been responsible for outcomes outside the United States that have not been described in another category?

Assessment

The EC should provide a qualitative assessment of the program and subprogram impact, discussing the evidence of reductions in injuries and illnesses or their appropriate proxies.

III.B.9. Review of Potential Outcomes

There may be health and safety impacts not yet appreciated and other beneficial social, economic, and environmental outcomes as a result of NIOSH activities. NIOSH study results may be influential outside the United States, and there may be evidence of implementation of NIOSH recommendations and training programs abroad.

Questions to Guide the Evaluation Committee

1. Is the program likely to produce a favorable change that has not yet occurred or not been appreciated?
2. Has the program been responsible for social, economic, security, or environmental outcomes?
3. Has the program's work had an impact on occupational health and safety in other countries?

Assessment

The EC may discuss other outcomes, including beneficial changes that have not yet occurred; social, economic, security, or environmental outcomes; and the impact that NIOSH has had on international occupational safety and health.

III.B.10. Summary Evaluation Ratings and Rationale

The EC should use its expert judgment to rate the relevance and impact of the overall research program by first summarizing its assessments of the major subprograms and then appropriately weighting the subprograms to determine the overall program ratings.

Table 5 provides some background context to aid the EC in reaching overall ratings for relevance and impact. The EC could consider the items in Table 5 for each subprogram then for the overall program and assess the relevance of the research subprograms and program by reviewing earlier responses to the questions in Sections III.B.2. through III.B.5. (reviews of program challenges, strategic goals and objectives, inputs, and activities). Items 1-4 in Table 5 are pertinent to assessing relevance.

To assess overall impact, the EC first needs to consider the available evidence of changes in work-related risks and adverse effects and external factors related to the changes. The EC should review the responses to the questions in Sections III.B.6. through III.B.8. (reviews of outputs, intermediate outcomes, and end outcomes) and systematically assess the impact of the research program and its subprograms. Items 5-7 in Table 5 will be helpful. The EC should evaluate separately the impact of the research and the impact of transfer activities. Transfer activities occur in two contexts: NIOSH efforts to translate intellectual products into practice and stakeholder efforts to integrate NIOSH results into the workplace. High impact assessments require the EC's judgment that the research program has contributed to outcomes; for example, outcomes have occurred earlier than they would have or are better than they would have been in the absence of the research program,

TABLE 5 Background Context for Program Relevance and Impact

Assess the following for each subprogram:

1. Relevance of current and recently completed research and transfer activities to objective improvements in workplace safety and health.
2. Contributions of NIOSH research and transfer activities to changes in work-related practices and reduction in workplace exposures, illnesses, or injuries.
3. Contributions of NIOSH research and transfer activities to improvements in work-related practices.
4. Contributions of NIOSH research to productivity, security, or environmental quality (beneficial side effects).
5. Evidence of reduction of risk in the workplace (intermediate outcome).
6. Evidence of reduction in workplace exposure, illness, or injuries (end outcome).
7. Evidence of external factors that prevented translation of NIOSH research results into intermediate or end outcomes.

or outcomes would have occurred were it not for external factors beyond NIOSH's control or ability to plan around.

The EC must assign one overall integer score for the *relevance* of the research program to the improvement of occupational safety and health and one overall integer score for the *impact* of the program on the improvement of occupational safety and health. The EC will use its expert judgment, summary assessment of research-program elements, and any appropriate information to arrive at those two scores. In light of substantial differences among the types of research programs that will be reviewed and the challenge to arrive at a summative evaluation of both relevance and impact, the FC chose not to construct an algorithm to produce the two final ratings.

Relevance and impact scores will be based on five-point categorical scales established by the FC (see Boxes 2 and 3) in which 1 is the lowest and 5 the highest rating. The FC has made an effort to establish mutually exclusive rating categories in the scales. When the basis of a rating fits more than one category, the highest applicable score should be assigned. It is up to the EC to determine how individual subprograms should influence final scores. Single integer values should be assigned. Final program ratings will consist of integer scores for relevance and impact and prose justification of the scores.

Box 2 includes the criteria for scoring the overall relevance of the NIOSH research program. As discussed in previous sections, numerous factors can be considered in assessing relevance. The scoring criteria focus on two: the EC assessment

BOX 2
Scoring Criteria for Relevance

5 = Research is in high-priority subject areas and NIOSH is significantly engaged in appropriate transfer activities for completed research projects/reported research results.
4 = Research is in priority subject areas and NIOSH is engaged in appropriate transfer activities for completed research projects/reported research results.
3 = Research is in high priority or priority subject areas, but NIOSH is not engaged in appropriate transfer activities; or research focuses on lesser priorities but NIOSH is engaged in appropriate transfer activities.
2 = Research program is focused on lesser priorities and NIOSH is not engaged in or planning some appropriate transfer activities.
1 = Research program is not focused on priorities and NIOSH is not engaged in transfer activities.

BOX 3
Scoring Criteria for Impact

5 = Research program has made major contribution(s) to worker health and safety on the basis of end outcomes or well-accepted intermediate outcomes.
4 = Research program has made some contributions to end outcomes or well-accepted intermediate outcomes.
3 = Research program activities are ongoing and outputs are produced that are likely to result in improvements in worker health and safety (with explanation of why not rated higher). Well accepted outcomes have not been recorded.
2 = Research program activities are ongoing and outputs are produced that may result in new knowledge or technology, but only limited application is expected. Well accepted outcomes have not been recorded.
1 = Research activities and outputs do not result in or are NOT likely to have any application.

of whether the program appropriately sets priorities among research needs and the EC assessment of how engaged the program is in appropriate transfer activities. Table 6 provides some guidance regarding how the EC may weight research priorities and transfer levels when determining relevance scores.

The EC will consider both completed research and research that is in progress and related to likely future improvements in its assessment of relevance. The EC should keep in mind how well the program has considered the frequency and severity of the problems being addressed; whether appropriate attention has been directed to both sexes, vulnerable populations, or hard-to-reach workplaces; and whether the different needs of large and small businesses have been accounted for. It is up to the EC to determine how to consider external factors in assigning program scores.

Box 3 includes the criteria established for the rating of impact. In general, the EC will consider completed research outputs during the assessment of impact. In assigning a score for impact, it is important to recognize that a "major contribution" (required for a score of 5) does not imply that the NIOSH program was solely responsible for observed improvements in worker health and safety. Many factors may be required to effect improvements. The EC could say that NIOSH made "major contributions" if the improvements would not have occurred when they did without NIOSH efforts.

TABLE 6 Guidance for Weighting Research Priority and Engagement in Appropriate Transfer Activities in the Application of Relevance Score

Assessment of Research Priority	Engagement in Applicable Transfer Activities	Applicable Score
High priority	Significantly engaged	5
High priority	Engaged	4
High priority	Not engaged	3
Priority	Significantly engaged	4
Priority	Engaged	4
Priority	Not engaged	3
Lesser priority	Significantly engaged	3
Lesser priority	Engaged	3
Lesser priority	Not engaged	2
Not focused on priorities	Significantly engaged	2
Not focused on priorities	Engaged	2
Not focused on priorities	Not engaged	1

The FC has some concern that the imposed scoring criteria for impact might be considered a promotion of the conventional occupational-health research paradigm that focuses on health-effects and technology research without much emphasis on the socioeconomic, policy, surveillance, and diffusion research (as opposed to diffusion activities) needed to effect change. The EC should remember that not all intermediate outcomes occur in the workplace. Important outcomes that NIOSH can effect also occur much farther out on the causal chain. NIOSH, for example, has an important role to play in generating knowledge that may contribute to changing norms in the insurance industry, in health-care practice, in public-health practice, and in the community at large. The EC may find that some of those issues need to be addressed and considered as external factors that facilitate or limit application of more traditional research findings. Given the rapidly changing nature of work and the workforce and some of the intractable problems in manufacturing, mining, and some other fields, the EC is encouraged to think beyond the traditional paradigm.

III.C. Assessment of NIOSH Process for Targeting Priority Research Needs and Committee Assessment of Emerging Issues

The second charge to the EC is the assessment of the research program's effectiveness in targeting new research and identifying emerging issues in occupational safety and health most relevant to future improvements in workplace protection. The EC is also asked to provide a qualitative narrative assessment of the program's efforts and to make suggestions about emerging issues that the program should be prepared to address. Among the most challenging aspects of research in illness and injury prevention are the identification of new or emerging needs or trends and the formulation of a research response that appropriately uses scarce resources in anticipation of them.

The EC should review the procedures that NIOSH and the research program have in place to identify needed research relevant to the NIOSH mission and should review the success that NIOSH has had in identifying and addressing research related to emerging issues. It should examine leading indicators from appropriate federal agencies, such as EPA, the Department of Labor, the National Institute of Standards and Technology, NIH, DOD, and the Department of Commerce. Those indicators should track new technologies, new products, new processes, and disease or injury trends.

One source of information deserving particular attention is NIOSH HHE reports. The HHE program offers a potential mechanism for identifying emerging research needs that could be incorporated as input into each of the programs evaluated. The EC should determine whether the program under review appropriately considers pertinent HHE investigation findings. Additional emerging issues may be revealed through consideration of NIOSH and the NIOSH-funded FACE reports, the AOEC reports, the U.S. Chemical Safety Board investigations, and SENSOR and other state-based surveillance programs. Appropriate federal advisory committees and other stakeholder groups should also be consulted to provide qualitative information.

The EC should systematically assess how the research program and its subprograms target new research by evaluating each subprogram for the items listed in Table 7. The EC will have to determine how best to weight subprogram contributions in the program's targeting of new research.

TABLE 7 Targeting of New Research and Identification of Emerging Issues

Assess the following for each subprogram:

1. Past and present effectiveness in targeting most relevant research needs.
2. Effectiveness in targeting research in fields most relevant to future improvements in occupational safety and health.
3. Contribution of NIOSH research to enhancement of capacity in government or other research institutions.

Questions to Guide the Evaluation Committee

1. What information does NIOSH review to identify emerging research needs?
    a. What is the process for review?
    b. How often does the process take place?
    c. How are NIOSH staff scientists and NIOSH leadership engaged?
    d. What is the process for moving from ideas to formal planning and resource allocation?
2. How are stakeholders involved?
    a. What advisory or stakeholder groups are asked to identify emerging research targets?
    b. How often are such groups consulted, and how are suggestions followed up?
3. What new research targets have been identified for future development in the program under evaluation?
    a. How were they identified?
    b. Were lessons that could help to identify other emerging issues learned?
    c. Does the EC agree with the issues identified and selected as important and with the NIOSH response, or were important issues overlooked?
    d. Is there evidence of unwise expenditure of resources on unimportant issues?

The EC members should use their expert judgment both to evaluate the emerging research targets identified by NIOSH and to provide recommendations to NIOSH regarding additional research that NIOSH has not yet identified. Recommendations should include a brief statement of their rationale.

IV. EVALUATION COMMITTEE REPORT TEMPLATE

Consistency and comparability among EC report formats is desirable, but the FC recognizes that each NIOSH research program is different and that each EC is independent. The outline provided in Box 4 flows from the FC's review of NIOSH's generalized logic model (Figure 1), the evaluation flowchart (Figure 2), and the assessment model described earlier in this document. The EC should feel free to use or adapt this outline as necessary when organizing its final report. The FC encourages each EC to look at prior EC reports for organizational ideas.

BOX 4
Suggested Outline for Evaluation Committee Reports

I. Introduction
This section should be a brief descriptive summary of the history of the program (and subprograms) being evaluated with respect to pre-NORA, NORA 1, and current and future plans of the research program presented by NIOSH. It should present the context for the research on safety and health; goals, objectives, and resources; groupings of subprograms; and any other important pertinent information. (A list of the NIOSH materials reviewed should be provided in Appendix C.)

II. Evaluation of Programs and Subprograms (Charge 1)
A. Evaluation summary (should include a brief summary of the evaluation with respect to impact and relevance, scores for impact and relevance, and summary statements).
B. Strategic goals and objectives: should describe assessment of the program and subprograms for relevance.
C. Review of inputs: should describe adequacy of inputs to achieve goals.
D. Review of activities: should describe assessment of the relevance of the activities.
E. Review of research-program outputs: should describe assessment of relevance and potential usefulness of the research program.
F. Review of intermediate outcomes and causal impact: should describe assessment of the intermediate outcomes and the attribution to NIOSH; should include the likely impacts and recent outcomes in the assessment.
G. Review of end outcomes: should describe the end outcomes related to health and safety and provides an assessment of the type and degree of attribution to NIOSH.
H. Review of other outcomes: should discuss health and safety impacts that have not yet occurred; beneficial social, economic, and environmental outcomes; and international dimensions and outcomes.
I. Summary of ratings and rationale.

III. NIOSH Targeting of New Research and Identification of Emerging Issues (Charge 2)
The EC should assess the progress that the NIOSH program has made in targeting new research in occupational safety and health. The EC should assess whether the NIOSH program has identified important emerging issues that appear especially important in terms of relevance to the mission of NIOSH. The EC should respond to NIOSH's perspective and add its own recommendations.

IV. Recommendations for Program Improvement
On the basis of the review and evaluation of the program, the EC may provide recommendations for improving the relevance of the NIOSH research program to health and safety conditions in the workplace and the impact of the research program on health and safety in the workplace.

Appendix A — Framework Document
Appendix B — Methods and Information-Gathering
Appendix C — List of NIOSH and Related Materials Collected in the Process of the Evaluation

B

Methods and Information Gathering

This appendix briefly describes sources of information used by the committee to perform its assessment of the National Institute for Occupational Safety and Health (NIOSH) Traumatic Injury (TI) Research Program's relevance and impact. Sources of information included an evidence package prepared by TI Research Program staff, stakeholder comment, and presentations by TI Research Program staff before the committee. A brief overview of the methods by which the committee evaluated information can be found in Chapter 1 of this report.

THE TI RESEARCH PROGRAM EVIDENCE PACKAGE

Information about the TI Research Program came in the form of an evidence package prepared by TI Research Program staff.[1] Each committee member was given a copy of the evidence package prior to the committee's first meeting. The evidence package included overviews of NIOSH and the TI Research Program as well as in-depth descriptions of the TI Research Program's work in each of the goal and subgoal areas. For each subgoal, there was an issue statement describing the research need that the TI Research Program is addressing, the approach (TI Research Program and extramural grantee overall strategy and research activities), outputs and transfers, and outcomes. For most subgoals there was a description of

---

[1]The evidence package is available for viewing on the NIOSH Web site at http://www.cdc.gov/niosh/nas/traumainj/.

external factors impacting the program and an overview of activities planned to extend research in the particular area. A draft of the TI Research Program's "Strategic Goals for the Future" was also included.

The evidence package included appendixes for each of the following: supporting evidence by goal or subgoal; TI intramural and extramural projects; TI management and research staff; TI informational resources; TI laboratory facilities and specialized equipment; TI partners and stakeholders; TI-sponsored or supported workshops and conferences; previous TI Program evaluations; and citation and dissemination data for TI output.

STAKEHOLDER COMMENT

NIOSH provided the Institute of Medicine (IOM) project staff with the names and contact information of 30 stakeholders identified as having an interest in occupational injury research and prevention. This list included representatives of associations, academia, product development, and federal and local government agencies. The list also included at least one representative for each of the eight traumatic injury research program goal areas (with the exception of motor vehicles), as well as four people identified as having a general interest in the TI Research Program.

Prior to the committee's final meeting, an e-mail inviting the aforementioned stakeholders to comment on the TI Research Program was prepared and sent to stakeholders by project staff on behalf of the committee. The e-mail explained the IOM committee's statement of task and provided an outline of the TI Research Program's goals and subgoals. Recipients were invited to submit comments on NIOSH's work in any or all of the goal areas. Additionally, recipients were encouraged to share the e-mail with other individuals or organizations with an interest in NIOSH's traumatic injury research. Stakeholders could submit comments by e-mail, fax, or a Web form.

Due to low response to the first e-mail, IOM project staff sent a follow-up e-mail after the committee's final meeting, reminding stakeholders of the opportunity to submit comments.

At the conclusion of the approximately 6-week comment period, the committee had received submissions from representatives of Femco, Inc. (a manufacturer of vehicle rollover protection devices); the National Truck Equipment Association (NTEA), Ambulance Manufacturers Division; Skyjack, Inc. (a manufacturer of fall protection equipment); the American Society of Safety Engineers (ASSE); and the Alaska Marine Safety Education Association (AMSEA). Those comments were then distributed to committee members for consideration in their evaluation of the TI Research Program. The AMSEA representative stated that NIOSH's research on Alaska's high-risk industries covers most areas of exposure in these occupations and did not provide any further comments. Comments from representatives of the other four organizations are summarized in Box B-1.

PRESENTATIONS BY TI RESEARCH PROGRAM STAFF BEFORE THE COMMITTEE

There were three in-person committee meetings, the first two of which included open sessions (see agendas below). The open portion of the committee's first meeting consisted of presentations by TI Research Program staff on the TI Research Program and each of its eight goals. A brief question-and-answer session followed each presentation.

During the open portion of the committee's second meeting, NIOSH staff gave presentations on NIOSH goal setting and prioritization as well as uses of surveillance data.

Open Session Agendas for First and Second Committee Meetings

Evaluation Committee Meeting One Presentations
by TI Program Staff—March 28, 2007

Charge to the Committee—NIOSH Overview
Lewis Wade, Ph.D.
NIOSH Senior Science Advisor

Overview of the Traumatic Injury Program
Nancy Stout, Ed.D.
Director, NIOSH Division of Safety Research
Program Manager, NIOSH Traumatic Injury Program

Reduce Injuries and Fatalities in the Transportation Sector
Stephanie Pratt, M.A.
Epidemiologist, NIOSH Division of Safety Research
Program Coordinator; Transportation, Warehousing, Utilities Sector Program

Reduce Injuries and Fatalities from Falls from Elevations
Hongwei Hsiao, Ph.D.
Chief, Protective Technology Branch
NIOSH Division of Safety Research

Reduce Injuries and Fatalities Due to Workplace Violence

Harlan Amandus, Ph.D.
Chief, Analysis and Field Evaluations Branch
NIOSH Division of Safety Research

Reduce Injuries and Fatalities Due to Machines

John Myers, M.S.F.
Health Statistician, NIOSH Division of Safety Research

Reduce Acute Back Injuries

James Collins, Ph.D.
Associate Director for Science
NIOSH Division of Safety Research

Reduce Injuries and Fatalities Among Workers in Alaska

Bradley Husberg, M.S.P.H.
Epidemiologist, NIOSH Alaska Field Station

Reduce Injuries and Fatalities to Emergency Responders

Timothy Pizatella, M.S.
Deputy Director, NIOSH Division of Safety Research

Reduce Injuries and Fatalities to Working Youth

Dawn Castillo, M.P.H.
Chief, Surveillance and Field Investigations Branch
NIOSH Division of Safety Research

Questions and Discussion

Evaluation Committee Meeting Two Presentations by TI Program Staff—May 31, 2007

Goal Setting and Prioritization

Nancy Stout, Ed.D.
Director, NIOSH Division of Safety Research
Program Manager, NIOSH Traumatic Injury Program

BOX B-1
Stakeholder Comments on NIOSH and the TI Research Program

Femco, Inc.—Rollovers on Tractors and Zero-Turn Radius Mowers

- Tractor and zero-turn radius mower accident reports are not widely available. Publication and dissemination of accident reports is needed to educate potential rollover protection structure (ROPS) users of the dangers of operating tractors and zero-turn radius mowers without ROPS.

NTEA, Ambulance Manufacturers Division—Improving Protection for Ambulance Workers in Patient Compartments

- Results of research to improve protection for ambulance workers in patient compartments (subgoal 7.1) should be officiated by the National Highway Traffic Safety Administration (NHTSA), as it relates to occupant protection in new motor vehicles and therefore falls under NHTSA's responsibility and authority.

Skyjack, Inc.—Fall Protection and Injuries and Fatalities Associated with Machinery

- Many injuries and fatalities involving the use of aerial lifts could be prevented through proper training of aerial lift operators, which would include coverage of aerial lift hazards, standards and regulations, and safe-use principles. Trainers should be certified compliant to the International Standardization Organization's standard 18878 of 2004,[a] which outlines methods to prepare training materials and to administer training to operators of mobile elevating work platforms.
- There has been and continues to be debate within the aerial lift industry with regard to proper use of fall protection on aerial lifts, in particular on scissor lifts and small personnel units. Research is needed to determine what fall protection uses are appropriate and safe for various types of aerial lifts.
- OSHA has fallen behind the industry with regard to maintaining regulations for aerial lift equipment. Many of the references within the regulations are outdated, and not specific to this equipment. Clearer, more focused regulations would eliminate confusion in the workplace, and ensure that employers are aware that appropriate training is required to operate these units.

ASSE

- ***Professionalism in Safety, Health, and Environmental (SH&E) Practice***

  NIOSH should support research to give both business and the safety and health community a better understanding of the training needed for SH&E professionals to

BOX B-1 *(continued)*

function appropriately as managers of workplace risks. Also, NIOSH should work with organizations offering safety certifications (e.g., the Board of Certified Safety Professionals, the American Board of Industrial Hygiene, and the Institute of Hazardous Material Management) to help determine how to apply what is known about safety and health professionals' competencies so that a comprehensive understanding of tasks and capabilities throughout the industry can be achieved. This same research can help examine other professional issues, such as:

–The appropriateness of SH&E education and training and how individuals enter and advance in the SH&E profession;

–The extent to which different SH&E professional segments have converged across traditional job roles;

–The role of technology in SH&E practice to advance effectiveness of SH&E management;

–The distribution of SH&E practitioners; and

–Ways to encourage more individuals to achieve the highest level of safety education, as well as formulate Ph.D. programs that will prepare individuals to meet future SH&E needs.

- ***NIOSH Involvement in Consensus Standards Process***

Increased NIOSH involvement in the national consensus standards process could help ensure that NIOSH research findings become operational in the field and could also bring NIOSH closer to safety and health practitioners and the challenges they face in advancing safety and health.

- ***Safety and Health Management Research***

NIOSH should sponsor more research that examines the role that broad safety and health management plays in the corporate and program structures of organizations. Greater commitment to safety and health at the highest management levels offers the best opportunity to reduce injuries, illnesses, and fatalities. Research is also needed to better understand how to make the business case for safety. NIOSH should be involved in bringing forth definitive, data-driven studies to help set a value on making a case for the workplace safety and health.

[a]ISO 18878: 2004—Mobile elevating work platforms—Operator (driver) training.

ACKNOWLEDGMENTS

The committee would like to thank the NIOSH staff for assembling the evidence package used by the committee for its evaluation. The committee would also like to thank the NIOSH staff, including Dr. Harlan Amandus, Ms. Dawn Castillo, Dr. James Collins, Dr. Honwei Hsiao, Mr. Bradley Husberg, Mr. John Myers, Mr. Timothy Pizatella, Ms. Stephanie Pratt, Dr. Nancy Stout, and Dr. Lewis Wade for their presentations and responses to questions from the committee and IOM staff at the committee's meetings. The committee also thanks Dr. Raymond Sinclair who served as the liaison between the committee and NIOSH. The committee especially extends its gratitude to Dr. Nancy Stout, who acted as the point of contact for the TI Research Program, and made time to respond to committee questions outside of meetings as well as to provide to the committee additional information not included in the evidence package.

The committee would also like to thank those who responded to the committee's request for stakeholder comment on the NIOSH TI Research Program. These were Mr. Brad Boehler of Skyjack, Inc., Ms. Kelly Comer of Femco, Inc., Mr. Jerry Dzugan of AMSEA, Mr. Steve Spata of NTEA, and Mr. Michael Thompson of ASSE.

The committee thanks the National Academies staff who provided assistance with the review, publication, and dissemination of this manuscript, including Lara Andersen, Clyde Behney, Bronwyn Schrecker Jamrok, Abbey Meltzer, Elisabeth Reese, and Jackie Turner. The committee also thanks for their guidance National Academies staff Andrew Pope, director of the IOM Board on Health Sciences Policy, as well as Evan Douple, Cathy Liverman, and Sammantha Magsino, staff to the Committee for the Review of NIOSH Research Programs. The committee thanks as well Rose Marie Martinez, director of the Board on Population Health, and Hope Hare for their support of this important project. The committee particularly thanks the study staff, Morgan Ford, Kathleen Stratton, and Kristina Van Doren Shulkin.

C

Information Provided by the NIOSH Traumatic Injury Research Program

Amandus, H. 2007. National Academies' Review of the NIOSH Traumatic Injuries Research Program. Goal 3: Reducing Occupational Injuries and Fatalities due to Workplace Violence. Presentation to the Committee to Review the NIOSH Traumatic Injury Research Program, Meeting I, March 30. Washington, DC.

Castillo, D. N. 2007. National Academies' Review of the NIOSH Traumatic Injuries Research Program. Goal 8: Reduce Injuries and Fatalities to Working Youth. Presentation to the Committee to Review the NIOSH Traumatic Injury Research Program, Meeting I, March 30. Washington, DC.

Collins, J. W. 2007. National Academies' Review of the NIOSH Traumatic Injuries Research Program. Goal 5: Reduce Acute Back Injury. Presentation to the Committee to Review the NIOSH Traumatic Injury Research Program, Meeting I, March 30. Washington, DC.

Hsaio, H. 2007. National Academies' Review of the NIOSH Traumatic Injuries Research Program. Goal 2: Reduce Injuries and Fatalities Due to Falls from Elevations. Presentation to the Committee to Review the NIOSH Traumatic Injury Research Program, Meeting I, March 30. Washington, DC.

Husberg, B. J. 2007. National Academies' Review of the NIOSH Traumatic Injuries Research Program. Goal 6: Reduce Injuries and Fatalities among Workers in Alaska. Presentation to the Committee to Review the NIOSH Traumatic Injury Research Program, Meeting I, March 30. Washington, DC.

Myers, J. R. 2007. National Academies' Review of the NIOSH Traumatic Injuries Research Program. Goal 4: Reduce Injuries and Fatalities Due to Machines. Presentation to the Committee to Review the NIOSH Traumatic Injury Research Program, Meeting I, March 30. Washington, DC.

NIOSH (National Institute for Occupational Safety and Health). 1998. Traumatic Occupational Injury Research Needs and Priorities: A Report by the NORA Traumatic Injury Team. Washington, DC: HHS.

NIOSH. 2007 (unpublished). Introduction (of evidence package). NIOSH.

NIOSH. 2007 (unpublished). Executive Summary (of evidence package). NIOSH.

NIOSH. 2007 (unpublished). Overview of the National Institute for Occupational Safety and Health. NIOSH.

NIOSH. 2007 (unpublished). Overview of Traumatic Injuries (TI) Research Program. NIOSH.

NIOSH. 2007 (unpublished). Current TI Research Goals and Sub Goals: Reduce Acute Back Injury. NIOSH.

NIOSH. 2007 (unpublished). Current TI Research Goals and Sub Goals: Reduce Injuries and Fatalities Among Workers in Alaska. NIOSH.

NIOSH. 2007 (unpublished). Current TI Research Goals and Sub Goals: Reduce Injuries and Fatalities Due to Falls from Elevations. NIOSH.

NIOSH. 2007 (unpublished). Current TI Research Goals and Sub Goals: Reduce Injuries and Fatalities Due to Machines. NIOSH.

NIOSH. 2007 (unpublished). Current TI Research Goals and Sub Goals: Reduce Injuries and Fatalities Due to Motor Vehicles. NIOSH.

NIOSH. 2007 (unpublished). Current TI Research Goals and Sub Goals: Reduce Injuries and Fatalities Due to Workplace Violence. NIOSH.

NIOSH. 2007 (unpublished). Current TI Research Goals and Sub Goals: Reduce Injuries and Fatalities to Emergency Responders. NIOSH.

NIOSH. 2007 (unpublished). Current TI Research Goals and Sub Goals: Reduce Injuries and Fatalities to Working Youth. NIOSH.

NIOSH. 2007 (unpublished). TI Strategic Goals for the Future. NIOSH.

NIOSH. 2007 (unpublished). Appendix 1: Supporting Evidence. NIOSH.

Pizatella, T. J. 2007. National Academies' Review of the NIOSH Traumatic Injuries Research Program. Goal 7: Reduce Injuries and Fatalities to Emergency Responders. Presentation to the Committee to Review the NIOSH Traumatic Injury Research Program, Meeting I, March 30. Washington, DC.

Pratt, S. 2007. National Academies' Review of the NIOSH Traumatic Injuries Research Program. Goal 1: Reduce Injuries and Fatalities Due to Motor Vehicles. Presentation to the Committee to Review the NIOSH Traumatic Injury Research Program, Meeting I, March 30. Washington, DC.

Stout, N. A. 2007. National Academies' Review of the NIOSH Traumatic Injuries Research Program. NIOSH Traumatic Injuries (TI) Research Program. Presentation to the Committee to Review the NIOSH Traumatic Injury Research Program, Meeting I, March 30. Washington, DC.

Stout, N. A. (NIOSH). 2007. Re: Information request from TI EC. E-mail to K. Stratton, Institute of Medicine. May 9.

Stout, N. A. (NIOSH). 2007. FW: Correction of Ongoing Intramural Project List. E-mail to K. Stratton, Institute of Medicine. May 11.

Stout, N. A. (NIOSH). 2007. TI Strategic Planning. E-mail to K. Stratton, Institute of Medicine. July 30.

Stout, N. A. (NIOSH). 2007. Injury Death Rates: Population vs. Occupational. E-mail to K. Stratton, Institute of Medicine. September 5.

Stout, N. A. (NIOSH). 2007. Trends in Workplace Homicides. E-mail to K. Stratton, Institute of Medicine. September 5.

Stout, N. A. (NIOSH). 2007. Revised Table 3 of TI-EP. E-mail to K. Stratton, Institute of Medicine. September 6.

Stout, N. A. (NIOSH). 2007. Research Project Planning and Resource Allocation. E-mail to M. Ford, Institute of Medicine. November 30.

Stout, N. A. (NIOSH). 2008. External Peer Review Process and Status of Alaska Field Station. E-mail to K. Stratton, Institute of Medicine. January 9.

Wade, L. V. 2007. National Academies' Review of the NIOSH Traumatic Injuries Research Program. Presentation to the Committee to Review the NIOSH Traumatic Injury Research Program, Meeting I, March 30. Washington, DC.

D

NIOSH TI Research Program Draft Strategic Goals for the Future

Strategic Goal 1: Reduce fall injuries in the workplace.

1.1: Reduce fall-related fatalities and injuries in the construction industry.

Intermediate Goal 1.1.1: Analyze existing data sets, including injury investigation reports, to identify the detailed sources and causes of falls-to-lower-level fatalities and injuries in the construction industry.

Intermediate Goal 1.1.2: Identify and summarize the existing measures, best practices, and intervention evaluations for preventing falls-to-lower-level in the construction industry for each injury source (i.e., roof, scaffold, and ladder) and priority task (i.e., roofing, masonry, painting and paper hanging, and carpentry).

Intermediate Goal 1.1.3: Identify and characterize falls from elevation among Hispanic construction workers.

Intermediate Goal 1.1.4: Evaluate the effectiveness of fall prevention technologies, programs, and communication materials in the construction industry.

Intermediate Goal 1.1.5: Transfer current knowledge and best practices on fall prevention and protection to construction special trade contractors, general building contractors, and Hispanic construction workers including current standards, PPE (personal protective equipment) technologies, and available technologies and methodologies in other fields that can be utilized in the construction applications.

Intermediate Goal 1.1.6: Provide the scientific basis for and facilitate the development of public health campaigns on fall prevention in the construction industry.

1.2: Reduce fall-related injuries in the health services industry.

Intermediate Goal 1.2.1: Characterize and track fall-related injuries in the health services industry, including determination of fall sources and causes.

Intermediate Goal 1.2.2: Identify and summarize preventive measures and products, current guidelines, best practices, and intervention evaluations for slips, trips, and falls (STFs) applicable to the health services industry.

Intermediate Goal 1.2.3: Evaluate the effectiveness of fall prevention technologies, programs, and communication materials in the health services industry.

Intermediate Goal 1.2.4: Summarize and transfer current knowledge on effective interventions and best practices for fall prevention and protection to nursing care facilities and hospitals as well as family service and residential care institutes.

Intermediate Goal 1.2.5: Provide the scientific basis for and facilitate the development of public health campaigns on fall prevention in the health services industry.

1.3: Reduce fall-related injuries in the wholesale and retail trade industry.

Intermediate Goal 1.3.1: Analyze existing data resources to determine the major subsectors in the retail and wholesale trade that have the highest fall-related injuries and fatalities and identify the sources and causes of fall injuries.

Intermediate Goal 1.3.2: Identify and summarize existing protective measures, best practices, and intervention evaluations for preventing falls in the high-risk subsectors of the retail and wholesale trade industry.

Intermediate Goal 1.3.3: Evaluate the effectiveness of fall prevention technologies, programs, and communication materials in the wholesale and trade industries.

Intermediate Goal 1.3.4: Summarize and transfer current knowledge on fall prevention and protection to retail merchandise stores that have the highest risk of fall-related injuries and fatalities (grocery stores and wholesale stores for vehicles, furniture, and construction materials suppliers).

Intermediate Goal 1.3.5: Provide the scientific basis for and facilitate the development of public health campaigns on fall prevention in the trade industry.

1.4: Reduce fall-related injuries through research on biosciences underlying human fall initiation, fall dynamics, fall termination, and control measures.

Intermediate Goal 1.4.1: Identify, analyze, and summarize the existing gaps in fall prevention and fall protection research and technology.

Intermediate Goal 1.4.2: Develop innovative fall prevention strategies that mitigate fall initiation hazards, especially during roofing, framing, scaffolding, and ladder-using jobs.

Intermediate Goal 1.4.3: Develop new and improved fall protection systems and devices to reduce the forces to the human body during fall arrest and fall termination.

Intermediate Goal 1.4.4: Develop scientifically comprehensive yet easy-to-use digital models of human fall dynamics, including the phases of fall initiation and fall termination, for use in efficient evaluation of new fall prevention and protection methods and strategies, in fall incident investigations (reconstruction), and in workers' hazard recognition training.

Strategic Goal 2: Reduce occupational injuries and deaths due to motor vehicles.

2.1: Reduce motor vehicle-related incidents in the TWU (transportation, warehousing, and utilities) industry sector. [TWU Strategic Goal 7]

Intermediate Goal 2.1.1: Characterize and track injuries among truck drivers. [TWU Strategic Goal 1]

Intermediate Goal 2.1.2: Identify risk factors associated with occupational motor vehicle crashes. [TWU Strategic Goal 7.2]

Intermediate Goal 2.1.3: Reduce motor vehicle-related injuries and deaths due to fatigue, distraction, and sleep disturbance. [TWU Strategic Goal 4]

Intermediate Goal 2.1.4: Evaluate the effectiveness of technologies and other means of reducing the incidence of motor vehicle-related injuries. [TWU Strategic Goal 7]

2.2: Reduce fatal and serious nonfatal injuries to workers in roadway construction work zones.

Intermediate Goal 2.2.1: Quantify worker exposure to being run over by construction vehicles and equipment operating inside roadway construction work zones. [Construction Intermediate Goal 2.1]

Intermediate Goal 2.2.2: Evaluate the potential for proximity warning system (PWS) use on road construction equipment to reduce worker exposure to being run over. Assess existing systems and explore development of new improved PWS technology. [Construction Intermediate Goal 2.2]

Intermediate Goal 2.2.3: Evaluate the potential for internal traffic control plans (ITCPs) to reduce worker exposure to being run over by construction equipment. [Construction Intermediate Goal 2.3]

Intermediate Goal 2.2.4: Improve availability and use of operator visibility limit information for road construction vehicles and equipment. [Construction Intermediate Goal 2.4]

Intermediate Goal 2.2.5: Evaluate worker injury risks associated with the expanded use of night work in the road construction industry. [Construction Intermediate Goal 2.5]

Intermediate Goal 2.2.6: Provide the scientific basis for campaigns to promote widespread usage of effective prevention measures in the road construction industry. [Construction Intermediate Goal 2.6]

2.3: Reduce injuries and fatalities from motor vehicle incidents, including being struck by vehicles, among public safety and emergency response workers. [Services Intermediate Goal 1]

Intermediate Goal 2.3.1: Identify and evaluate the leading risk factors for motor-vehicle incidents among public safety and emergency response workers, including being struck by vehicles, and provide effective recommendations to reduce the number of fatalities and days-away-from-work incidents. [Services Intermediate Goal 1]

Intermediate Goal 2.3.2: Identify and evaluate the effectiveness of interventions such as driver training in reducing motor vehicle-related injuries and deaths among public safety and emergency response workers, including volunteer response workers such as volunteer firefighters. [Services Intermediate Goal 1]

Intermediate Goal 2.3.3: Develop and promulgate information on appropriate body restraints for firefighters and emergency medical service when operating or riding in emergency response vehicles. [Services Intermediate Goal 1]

2.4: Reduce occupational road traffic injuries worldwide. [TWU Strategic Goal]

Intermediate Goal 2.4.1: Identify risk factors for occupational road traffic injuries outside the United States.

Intermediate Goal 2.4.2: Develop road safety materials targeted to employees posted overseas.

Strategic Goal 3: Reduce occupational injuries and deaths due to workplace violence.

3.1: Reduce workplace violence in the transportation, warehouse, and utilities industries.

Intermediate Goal 3.1.1: Develop and assess models that increase participation and compliance to effective interventions that reduce violence among high-risk taxicab driver populations.

3.1.1.1: Conduct research studies to demonstrate the effectiveness of model programs that increase compliance of high-risk taxicab drivers to adopt effective interventions that reduce risk of violence.

3.1.1.2: Develop partnerships with community organizations, police departments and their associations, and taxicab associations, and implement a model program in at least one community to increase the number of taxicabs compliant with effective interventions.

3.2: Reduce workplace violence among high-risk wholesale and retail trade workers including grocery stores, gasoline stations, convenience stores, bakeries, and liquor stores.

Intermediate Goal 3.2.1: Increase the reliability and availability of surveillance data sets for workplace violence in high-risk wholesale and retail trade industries and occupations.

Intermediate Goal 3.2.2: Identify factors that increase the risk of workplace violence among wholesale and retail workers who are employed in high-risk wholesale and retail establishments.

Intermediate Goal 3.2.3: Implement effective interventions that reduce violence among high-risk populations of workers in the high-risk wholesale and retail trade sector.

3.2.3.1: Identify management practices that minimize the risks of workplace violence in high-risk wholesale and retail establishments.

3.2.3.2: Complete an assessment of intervention methods to reduce the incidence and severity of violent incidents.

3.2.3.3: Identify and evaluate successful community policing program models, which increases participation and compliance to implementing NIOSH and OSHA guidelines for prevention of robbery and robbery-related assault in late night retail businesses.

Intermediate Goal 3.2.4: Improve transfer, diffusion, and adoption of effective workplace violence interventions in high-robbery-risk wholesale and retail trade sector businesses.

3.2.4.1: Utilize partnerships with employers, labor unions, trade associations, police departments and their associations, and government agencies to increase the knowledge of the risks of workplace violence and potential strategies and interventions that limit risks in high-risk wholesale and retail trade businesses.

3.2.4.2: Develop and implement communication plans for effective workplace violence intervention strategies that are demonstrated to have reached target audiences in the wholesale and retail trade sector.

3.2.4.3: Utilize partnerships with employers, trade associations, government agencies, police departments and their associations, and crime prevention organizations to increase knowledge of workplace violence prevention among police departments, and implement a successful community policing program that increases business compliance with interventions and reduces robbery-related assaults in three communities.

Intermediate Goal 3.2.5: Use reliable economic models to accurately assess the costs of fatal and nonfatal injuries from workplace violence and the potential savings from reducing incidence and severity of workplace violence events.

3.2.5.1: Complete an assessment of reduction in costs due to implementation of a successful community policing program in one community.

3.2.5.2: Disseminate return on investment and cost data to community partners to promote compliance with NIOSH and OSHA guidelines for retail workplace violence prevention.

3.3: Identify risk factors and effective interventions to prevent workplace violence among high-risk services, healthcare, and social service sector workers such as eating and drinking establishment workers; hotels or motel workers; automotive repair mechanics; teachers; nurses and nursing assistants in general medical, home health care, nursing homes, and psychiatric hospitals; social service workers in job training, residential care, and day care industries; private security workers; and public safety and correctional workers in emergency response tasks (e.g., medical services, police calls, and correctional officer activities).

Intermediate Goal 3.3.1: Conduct studies to identify factors associated with trauma from workplace violence among high-risk service sector workers.

3.3.1.1: Conduct etiologic research studies to identify the conditions that lead to physical or psychological trauma from workplace violence among high-risk service sector workers.

3.3.1.2: Identify work organization strategies that minimize the risks of psychological trauma due to harassment, intimidation, bullying, and sabotage.

3.3.1.3: Evaluate the factors associated with fatal and nonfatal workplace violence incidents among youth worker populations in high-risk service sector industries.

Intermediate Goal 3.3.2: Conduct studies to increase the reliability and availability of surveillance data sets for workplace violence in high-risk service sector industries and occupations.

3.3.2.1: Identify a prioritized list of essential information for efficient population-based surveillance of verbal aggression and physical trauma from workplace violence in high-risk industries and occupations and in priority populations (similar to data collected through the National Crime Victimization Survey for physical trauma).

3.3.2.2: Develop at least one effective partnership with employers, unions, government, or nongovernmental organizations to collect ongoing surveillance information about verbal aggression and physical trauma from workplace violence.

3.3.2.3: Interpret surveillance information to help identify and eliminate workplace conditions that lead to physical trauma from workplace violence, and disseminate the results.

3.3.2.4: Distribute and publicize surveillance data sets that include parameters on workplace violence for widespread use by extramural investigators.

Intermediate Goal 3.3.3: Develop effective interventions that reduce violence among high-risk populations of workers in the service sector.

3.3.3.1: Validate at least three intervention methods to reduce the incidence and severity of violence in high-risk service sector industries.

3.3.3.2: Develop at least three partnerships with employers or trade associations in high-risk service sector industries to implement interventions in workplace violence programs and evaluate compliance with effective intervention strategies that have been demonstrated to reduce workplace violence.

Intermediate Goal 3.3.4: Improve transfer, diffusion, and adoption of effective workplace violence interventions in high-risk service sector industries, particularly for small businesses.

3.3.4.1: Utilize partnerships with employers, labor unions, trade associations, and government agencies in the service sector to increase the knowledge of workplace violence risks and potential strategies to limit risks.

3.3.4.2: Develop and implement communication plans for effective workplace violence intervention strategies that are demonstrated to have reached target audiences in the range of small to large employers.

Intermediate Goal 3.3.5: Use reliable economic models to accurately assess the costs of fatal and nonfatal injuries from workplace violence and the potential savings from reducing the incidence and severity of workplace violence among service sector workers.

3.3.5.1: Disseminate return on investment and cost data to community partners to promote compliance with NIOSH and OSHA guidelines for retail workplace violence prevention.

Strategic Goal 4: Reduce occupational injuries and deaths due to machines and industrial vehicles.

4.1: Reduce occupational injuries and deaths due to machines and industrial vehicles in the agriculture, forestry, and fishing industry, with an emphasis on tractor-related injuries and deaths.

Intermediate Goal 4.1.1: Complete testing and market development of the NIOSH AutoROPS for zero-turn riding mowers for adoption by manufacturers.

Intermediate Goal 4.1.2: Complete testing and market development of the NIOSH AutoROPS for medium-horsepower tractors for adoption by manufacturers.

Intermediate Goal 4.1.3: Complete testing and market development of the NIOSH cost-effective rollover protective structure (CROPS) for retrofitting older tractors without ROPS for adoption by manufacturers.

Intermediate Goal 4.1.4: Characterize and track injuries and fatalities from machinery and industrial vehicles in commercial fishing.

Intermediate Goal 4.1.5: Identify and prioritize risk factors associated with machinery and industrial vehicles in commercial fishing.

Intermediate Goal 4.1.6: Reduce the incidence of on-deck machinery entanglements on commercial fishing vessels through continued development of emergency-stop (e-stop) and other machine guarding technologies.

Intermediate Goal 4.1.7: Characterize and track injuries and fatalities from machinery and industrial vehicles in logging.

Intermediate Goal 4.1.8: Identify and prioritize risk factors associated with machinery and industrial vehicles in logging.

Intermediate Goal 4.1.9: Identify work situations at high risk for machine-related fatal injury, and develop prevention strategies for those who can intervene in the workplace by conducting Fatality Assessment and Control Evaluation (FACE) investigations of targeted occupational fatality incidents.

4.2: Reduce occupational injuries and deaths due to machines and industrial vehicles in the construction industry.

Intermediate Goal 4.2.1: Quantify worker exposure to being run over by construction vehicles and equipment operating inside roadway construction work zones. [Construction Intermediate Goal 2.1]

Intermediate Goal 4.2.2: Evaluate the potential for proximity warning system (PWS) on road construction equipment to reduce worker exposures to being run over. Assess existing systems and explore development of new improved PWS technology. [Construction Intermediate Goal 2.2]

Intermediate Goal 4.2.3: Evaluate the potential for internal traffic control plans (ITCPs) to reduce worker exposure to being run over by construction equipment. [Construction Intermediate Goal 2.3]

Intermediate Goal 4.2.4: Improve availability and use of operator visibility limit information for road construction equipment. [Construction Intermediate Goal 2.4]

Intermediate Goal 4.2.5: Evaluate risks of worker injury associated with the expanded use of night work in the road construction industry. [Construction Intermediate Goal 2.5]

Intermediate Goal 4.2.6: Provide the scientific basis for campaigns to promote widespread use of effective prevention measures in the road construction industry.

Intermediate Goal 4.2.7: Validate a computer simulation model to analyze the impact of dynamic loading of scissor lifts for evaluating fall, collapse, and tip-over incidents for use in improving equipment design and developing effective safety devices for adoption by equipment manufacturers.

Intermediate Goal 4.2.8: Identify work situations at high risk for machine-related fatal injury and develop prevention strategies for those who can intervene in the workplace by conducting Fatality Assessment and Control Evaluation (FACE) investigations of targeted occupational fatality incidents.

4.3: Reduce occupational injuries and deaths due to machines and industrial vehicles in the manufacturing industry.

Intermediate Goal 4.3.1: Characterize and track injuries and fatalities from machinery and industrial vehicles.

Intermediate Goal 4.3.2: Identify and prioritize risk factors associated with machinery and industrial vehicles.

Intermediate Goal 4.3.3: Identify work situations at high risk for machine-related fatal injury, and develop prevention strategies for those who can intervene in the workplace by conducting Fatality Assessment and Control Evaluation (FACE) investigations of targeted occupational fatality incidents.

Intermediate Goal 4.3.4: Evaluate the effectiveness of technologies and other means of reducing the incidence of injuries and fatalities from high-risk machinery and industrial vehicles.

Intermediate Goal 4.3.5: Complete the testing and market development of the NIOSH JamAlert for passively controlling hazardous energy during unjamming tasks on industrial balers for adoption by manufacturers.

Intermediate Goal 4.3.6: Complete the NIOSH evaluation of the American Standards Institute (ANSI) B11 TR3 machine risk reduction methodology and provide recommendations to OSHA on the adoption of new ANSI and International Organization for Standardization (ISO) methods for machine risk reduction.

Intermediate Goal 4.3.7: Complete testing and market development of the NIOSH radio-frequency (HASARD) device for passively controlling hazardous energy from worker proximity to conveyors for adoption by manufacturers.

4.4: Reduce occupational injuries and deaths due to machines and industrial vehicles in the mining industry.

Intermediate Goal 4.4.1: Develop interventions for preventing injuries related to machine safety and powered haulage equipment. [Mining Intermediate Goal 4.2]

Intermediate Goal 4.4.2: Investigate wearable sensor technologies for empowering the miner to take proactive steps in decreasing his or her exposure to work-related injuries. [Mining Intermediate Goal 4.3]

Intermediate Goal 4.4.3: Develop interventions, best practices, and strategies for improving miners' training with respect to hazard recognition, risk factor awareness, and emergency response. [Mining Intermediate Goal 4.5]

Strategic Goal 5: Reduce occupational injuries and deaths among high-risk and vulnerable worker groups.

5.1: Reduce occupational injuries and deaths among youth workers.

Intermediate Goal 5.1.1: Track trends in youth worker injuries and deaths.

Intermediate Goal 5.1.2: Partner with educational groups to increase youth education on occupational safety.

Intermediate Goal 5.1.3: Provide empirical data to guide the development and enforcement of child labor laws.

Intermediate Goal 5.1.4: Partner with governmental agencies, child safety advocates, safety groups, trade associations, and others to communicate young worker safety information.

5.2: Reduce occupational injuries and deaths among older workers.

Intermediate Goal 5.2.1: Characterize and track trends in older worker injuries and deaths.

Intermediate Goal 5.2.2: Identify risk factors for injury and death among older workers.

Intermediate Goal 5.2.3: Evaluate intervention strategies to prevent older worker injuries and deaths.

Intermediate Goal 5.2.4: Partner with government agencies, senior citizen organizations, and others to communicate older worker safety information.

5.3: Reduce occupational injuries and deaths among high-risk ethnic and minority workers.

Intermediate Goal 5.3.1: Characterize and track trends in injuries and death among ethnic and minority workers.

Intermediate Goal 5.3.2: Identify risk factors for injury and death among high-risk ethnic and minority workers.

Intermediate Goal 5.3.3: Partner with government agencies, safety groups, environmental justice organizations, social organizations, and others to develop and evaluate effective, culturally sensitive prevention strategies.

Intermediate Goal 5.3.4: Partner with government agencies, safety groups, environmental justice organizations, social organizations, and others to communicate safety information to employers and high-risk ethnic and minority workers.

5.4: Reduce occupational injuries and deaths among immigrant workers.

Intermediate Goal 5.4.1: Characterize and track trends in injuries and death among immigrant workers.

Intermediate Goal 5.4.2: Identify risk factors associated with injury and death among immigrant workers.

Intermediate Goal 5.4.3: Partner with government agencies, safety professionals, immigrant organizations, and international agencies to communicate safety information to employers and immigrant workers.

E

Biographical Sketches of Committee Members

Brian L. Strom, M.D., M.P.H. (IOM) (*Chair*), is associate vice dean, chair, and professor in the Department of Biostatistics and Epidemiology, as well as a George S. Pepper Professor of Public Health and Preventive Medicine, professor of medicine, professor of pharmacology, and director of the Center for Clinical Epidemiology and Biostatistics at the University of Pennsylvania School of Medicine. Internationally known for multiple areas of clinical epidemiology, Dr. Strom's major career interest is pharmacoepidemiology, specifically looking at adverse drug reactions and medical errors. He received his medical degree from Johns Hopkins University School of Medicine and his master's of public health from the University of California at Berkeley. He is a member of the Institute of Medicine (IOM). He was chair of the IOM's Committee to Assess the Safety and Efficacy of the Anthrax Vaccine and the IOM's Committee on the Smallpox Vaccination Program Implementation, as well as a member of the IOM's Committee to Review the CDC (Centers for Disease Control and Prevention) Anthrax Vaccine Safety and Efficacy Research Program.

Robin Baker, M.P.H., is the coordinator of public programs at the University of California (UC) at Berkeley's School of Public Health and director of the Labor Occupational Health Program (LOHP). LOHP is a community service program that provides training, information, and assistance to workers, labor organizations, joint labor-management health and safety committees, community-based environmental justice groups, and others. Robin Baker is a health educator with more than 20 years of experience in the occupational health field and is an instructor in the graduate public health education program at UC Berkeley. She has published numerous articles and resources on worker training. She currently directs more than a dozen federal- and state-funded projects ranging from an examination of the experience of injured workers in the California Workers' Compensation system to a pilot school-based program to educate teen workers. Before coming to LOHP in 1981, she directed a worker training program for electronics workers in Silicon Valley and served on the staff of the California-Occupational Safety and Health Administration (Cal/OSHA) education unit. In addition, she serves as director of several major projects, including the California Worker Occupational Safety and Health Training and Education Program, the Working Immigrant Safety and Health (WISH) Coalition, and the National Young Worker Health and Safety Resource Center.

Leslie I. Boden, Ph.D., is professor of public health and associate chair of the Department of Environmental Health at Boston University School of Public Health (BUSPH). He is an economist, and much of his research has focused on finding ways to highlight the economic and human consequences of injuries and illnesses and to identify ways of minimizing those consequences. Over the past several years, Dr. Boden has published studies measuring the income lost by injured workers and the adequacy of workers' compensation benefits. With BUSPH colleague Lee Strunin, he has also published several studies of the post-injury experiences of workers and their families. Recently, Dr. Boden has been working on a study that estimates underreporting of workplace injuries. He has also written on occupational safety and health regulation, medical screening, gender inequality, and the legal and public health use of scientific information. From 1988 to 1997, Dr. Boden served on the Mine Health Research Advisory Committee of the Department of Health and Human Services, which he chaired for 6 years. In 2001-2002, he was a member of the Worker Advocacy Advisory Group, which advised the Department of Energy on occupational disease compensation. He has also co-chaired a group advising the National Institute for Occupational Safety and Health (NIOSH) on its research agenda on the social and economic consequences of workplace illness and injury.

Barry Bozeman, Ph.D., is an Ander Crenshaw Professor and Regents' Professor of Public Policy at the University of Georgia. He holds an appointment as adjunct honorary professor of political science at the University of Copenhagen. Before joining the University of Georgia he was Regent's Professor of Public Policy, Georgia Tech, and professor of public administration and adjunct professor of engineering, the Maxwell School, Syracuse University. At Georgia Tech, he was the first full-time director of the School of Public Policy and the founding director of the Research

Value Mapping Program. At Syracuse University, he was founding director of the Center for Technology and Information Policy. Bozeman's practitioner experience includes a position at the National Science Foundation's Division of Information Technology and a visiting position at the Science and Technology Agency's (Japan) National Institute of Science and Technology Policy. Bozeman is coeditor of the *Journal of Technology Transfer*. He has served as a consultant to a variety of federal and state agencies in the United States, including the Internal Revenue Service, the Department of Commerce, the National Science Foundation (NSF), and the Department of Energy (DOE). He has helped in the design and evaluation of the national innovation systems of the Republic of South Africa, Canada, New Zealand, France, Chile, and Argentina. His research has been funded by grants from NSF, DOE, the National Institutes of Health (NIH), the Department of Commerce, the Environmental Protection Agency (EPA), the Office of Naval Research, the Kellogg Foundation, the Sloan Foundation, and the Rockefeller Foundation, among others. He has served on four National Academy of Sciences-National Academy of Engineering panels. He is author or editor of 15 books.

Stephen W. Hargarten, M.D., M.P.H., is professor and chairman of the Department of Emergency Medicine at the Medical College of Wisconsin (MCW). He has been practicing emergency medicine for more than 30 years. He received his M.D. from the Medical College of Wisconsin and his M.P.H. from the Bloomberg School of Public Health. He currently serves as the director of the Injury Research Center at MCW, 1 of 12 CDC-funded centers in the United States. For more than 20 years he has been an active injury prevention and control researcher, educator, and practitioner. He recently served on the advisory committee for developing the acute care research agenda for the National Injury Control and Prevention Center of CDC. Dr. Hargarten has published frequently in the peer-reviewed literature on injury topics. His major injury-related activities include injury surveillance (particularly linked data sets), violence, injury of travelers, and trauma system development. He currently also serves as the director of the Firearm Injury Center at MCW. He is the immediate past president of the Society for Advancement of Violence and Injury Research and is currently a member of the board of Advocates for Highway and Auto Safety.

Brian M. Kleiner, Ph.D., is professor of industrial and systems engineering and founding director of the Center for Innovation in Construction Safety and Health, a core research unit in the Myers-Lawson School of Construction at Virginia Polytechnic Institute and State University (Virginia Tech). He is immediate past director of the Human Factors Engineering and Ergonomics Center and currently directs the Macroergonomics and Group Decision Systems Laboratory in the Grado Department of Industrial and Systems Engineering. His interest in the sociotechnical nature of work systems is supported by his educational training in both psychology and engineering. He received his M.S. and Ph.D. in industrial engineering (human factors engineering-ergonomics concentration) from the State University of New York at Buffalo and has a B.A. in psychology. Dr. Kleiner's research interests focus on systems or macroergonomics for improved safety, health, and performance.

Tom B. Leamon, Ph.D., C.P.E., was a vice president of Liberty Mutual Insurance Group and is director emeritus of the Liberty Mutual Research Institute for Safety. Dr. Leamon was responsible for the research program in occupational safety and rehabilitation and has published research on integrated approaches to occupational injury and illness, industrial ergonomics, and evaluation of criteria for the prevention of low back pain disability. He is currently developing a significant surveillance study of occupational injury in Vietnam. He also serves as an adjunct professor of occupational safety at the Harvard School of Public Health and is a term member of the NIOSH-National Occupational Research Agenda (NORA) liaison committee. Dr. Leamon is a fellow of the Ergonomics Society, the Human Factors and Ergonomics Society, and the Institution of Electrical Engineers (U.K.). He is a board-certified professional ergonomist, a chartered engineer (U.K.), and a European engineer. Dr. Leamon received his Ph.D. in industrial engineering from the Institute of Technology, Cranfield.

James M. Melius, M.D., Dr.P.H., is director of the New York State Laborers' Health and Safety Trust Fund and director of research for the Laborers' Health and Safety Fund of North America. His work focuses on the development and promotion of health and safety programs in the construction industry. He is currently a member of the Advisory Board on Radiation and Worker Health and is past chair of the Board of Scientific Counselors for the Agency for Toxic Substances and Disease Registries. He was also a member of the NIOSH Board of Scientific Counselors and of the CDC Advisory Committee for the Elimination of Tuberculosis. He worked for NIOSH from 1980 to 1987 where he directed NIOSH's main workplace consultation program. From 1987 to 1994, he worked for the New York State Department of Health where he directed environmental and occupational health programs including the development of a network of occupational health clinics.

Mark S. Redfern, Ph.D., is co-director of the Medical Virtual Reality Center as well as a William Kepler Whiteford Professor of Bioengineering at the University of Pittsburgh. He has appointments in the Departments of Bioengineering, Otolaryngology, and Rehabilitation Science. He received his B.S. in engineering science from the University of Michigan, Ann Arbor, in 1978. In 1980, he earned a certificate in prosthetics from New York University and worked for 3 years as a clinical prosthetist. In 1988 he earned a Ph.D. in bioengineering from the University of Michigan. At Pittsburgh, Dr. Redfern's research is focused in two areas: human postural control and ergonomics. The major goal of the postural control research is the prevention of falling injuries by investigating the factors that influence balance in older adults and patients with balance disorders. The ergonomic research focus has been in the area of fall prevention and reducing injuries in the workplace through ergonomic redesign.

Gordon R. Reeve, Ph.D., recently retired from the position of corporate epidemiologist at Ford Motor Company after having spent 21 years in the public health industry. He received his Ph.D. from the University of Texas School of Public Health in 1982, with major concentrations in epidemiology and toxicology as well as industrial hygiene. He also received a master's degree in public health from the University of Texas in 1975, where his focus was occupational health. From 1975 through 1979, he coordinated a program at the University of Texas Cancer Center to evaluate the effectiveness of lung cancer screening for asbestos workers. In the summer of 1979 he began work at NIOSH in Cincinnati. During the next 5 years, he conducted several occupational cancer studies involving chemical industry workers, as well as a major study of heart disease among munitions workers exposed to nitroglycerin. In 1989, Dr. Reeve obtained the position of corporate epidemiologist at Ford Motor Company. At Ford, he conducted occupational health research jointly with the United Auto Workers Union, which involved 800,000 of Ford's current and former employees. Dr. Reeve completed the major effort of designing a near-real-time injury surveillance system, which was installed at 57 Ford plant locations in 1994. His research interests and assignments include risk factors associated with acute and cumulative trauma, linkages between injury data and manufacturing quality indicators, and patterns of healthcare utilization.

Joseph J. Schwerha, M.D., M.P.H., is professor of occupational medicine in the Graduate School of Public Health at the University of Pittsburgh and is the director of the Occupational and Environmental Medicine Residency and the Public Health Preparedness and Disaster Response Certificate Programs. Prior to working at the University of Pittsburgh, Dr. Schwerha was general manager of health services and corporate medical director at U.S. Steel. His research interests are in medical administration and education as well as all aspects of environmental health and safety and medical surveillance. He serves on the IOM Standing Committee on Personal Protective Equipment and the Committee on Training Physicians for Public Health Careers. He holds a bachelor's degree in chemistry from the University of Pittsburgh, a master's of public health in environmental health and industrial hygiene from the University of Michigan, and a medical degree from West Virginia University School of Medicine. He is actively involved on the editorial boards of the *Journal of Occupational and Environmental Medicine* and the *Official Disability Guidelines.* He is a recipient of the William S. Knudsen Award, the highest international award in occupational medicine, from the American College of Occupational and Environmental Medicine in 2005.

FRAMEWORK COMMITTEE LIAISON

Letitia K. Davis, Sc.D., Ed.M., is director of the Occupational Health Surveillance Program in the Massachusetts Department of Public Health where she has worked for the last 20 years to develop state-based surveillance systems for work-related illnesses and injuries. The Occupational Health Surveillance Program uses surveillance findings to inform state and local prevention activities and over the years has undertaken a variety of educational intervention activities, as well. Dr. Davis has served as principal investigator for a community-based intervention project to enhance the health and safety of young workers. More recently she has served as principal investigator for a study examining the feasibility of working with community health centers to collect data on occupational health experiences of immigrant workers. Dr. Davis is also lead occupational health consultant to the Council of State and Territorial Epidemiologists. Dr. Davis serves as adjunct faculty of the Department of Work Environment at the University of Massachusetts at Lowell and as a visiting lecturer on occupational health at the Harvard School of Public Health. She is also a past member of the NIOSH Board of Scientific Counselors and of the National Advisory Committee on Occupational Safety and Health. Dr. Davis received her doctorate in occupational health from the Harvard School of Public Health. She has previously served on the IOM Panel on the Health and Safety Implications of Child Labor.